About the Institute

The National Institute of Agricultural Extension Management, (MANAGE) at Hyderabad, is an apex Institute set up under the Ministry of Agriculture, Government of India in 1987. MANAGE has a mission to facilitate the Extension Officers, Managers, Scientists and Administrators in all sectors of the agricultural economy to acquire managerial and Technical skills that would enable them to provide most effective support and services to farmers and fishermen for practicing Sustainable Agriculture.

NATIONAL INSTITUTE OF AGRICULTURAL EXTENSION MANAGEMENT (MANAGE)

Rajendranagar,
Hyderabad - 500 030

Tel.: +91 (0) 40 4016702-706

Fax: +91 (0) 40 4015388

Email: MANAGE @ HUB.NIC.IN

FARMERS' PARTICIPATION IN AGRICULTURAL RESEARCH AND EXTENSION SYSTEMS

CONSULTANTS

K.C. JOHN
D.K. SHARMA

C.S. RAJAN, *IAS*
CHARANJIT SINGH

COORDINATOR

S.K. ARORA, *IAS*

Published for

NATIONAL INSTITUTE OF AGRICULTURAL EXTENSION MANAGEMENT, HYDERABAD

by

CONCEPT PUBLISHING COMPANY, NEW DELHI-110059

ISBN 81-7022-617-1

First Published 1997

Published and Printed by

Ashok Kumar Mittal

Concept Publishing Company

A/15-16, Commercial Block, Mohan Garden
NEW DELHI-110059

Phones: 5554042, 5504042
Fax 091-011-5598891

ACKNOWLEDGEMENTS

The preparation of the Report benefited greatly from the valuable contributions by numerous organizations and individuals.

Several groups of farmers from Andhra Pradesh, Assam, Gujarat, Karnataka, Maharashtra, Meghalaya, Rajasthan, Tamilnadu, Uttar Pradesh and West Bengal shared their rich experience and insights with the team. Their contribution and creativity have been a source of great encouragement and inspiration throughout the mission.

Researchers positioned in Zonal Research Stations and extension professionals in the above States participated in a series of brainstorm sessions. Vice-Chancellors, Directors of Research and Extension of State Agricultural Universities in Andhra Pradesh, Assam, Maharashtra, Tamilnadu and Uttar Pradesh collaborated with the team. Secretary, Department of Agriculture, Government of Meghalaya, Uttar Pradesh and West Bengal provided their assessments of the public extension and research systems.

Directors of Agriculture in the States that the team visited took care of logistic support and also provided their valuable suggestions.

The authors of the Report would also like to acknowledge with great appreciation the intellectual advice and encouragement from the National Steering Committee and the participants of the two National Workshops.

Secretarial and Administrative Support for the Report preparation was overseen by Mr. M. B. Salim, Administrative Officer, MANAGE. A team of secretarial staff assisted in the word processing of the manuscripts.

Authors

CONTENTS

Page

Acknowledgements 5
Executive Summary 9

1. Introduction 15
2. Methodology 21
3. Diagnosis 26
4. Agriculture Technology Development and Dissemination System 40
5. Recommendations 63
6. Pilot Projects on Farmer Participatory Research and Extension Systems in Irrigated Areas and Rainfed Farming Systems 110

ANNEXURES

I. Profile of Consultants 129
II. Terms of Reference of the Study on Introducing Farmer Participation into Planning and Managing Extension and Research Systems in India 132
III. Desk Review: National Institute of Agricultural Extension Management, Hyderabad 137
IV. Illustrative set of issues for facilitating discussions in the National Workshop held on 1-2 February, 1993, at MANAGE, Hyderabad 167
V. Proceedings of the National Workshop held on 1-2 February 1993, at MANAGE, Hyderabad 170
VI. Report on Field Visits to Farmer Participatory Projects 173
VII. Guidelines on Field Visits to Different Agricultural Situations 213

		Page
VIII.	**Proceedings of Second National Workshop**	222
IX.	**Farmer Participatory Study: Draft Recommendations**	233
X.	**Itinerary of Consultants**	259
	Bibliography	264
	Index	267

EXECUTIVE SUMMARY

1. There is an increasing recognition in India that the agricultural extension and research systems, while responding to new and emerging challenges, should become more farmer-relevant and accountable to their clients. This report examines whether the extension and research systems, which have had an impeccable track record in the past, can adopt farmer participatory approaches.

2. The report explores issues like the reorganization of institutional processes to make them flexible and responsive, approaches and methods that encourage functional participation vis-a-vis empowerment of farmer organizations, and the need to embed the reforms into the extension and research systems management.

3. In well endowed regions, advantages of genetic gains, fertilizer responsiveness and assured moisture regime have been exploited to a large extent. Second generation problems, such as nutrient mining, draw down, pathogen build up, deterioration in soil quality, dynamic changes in the weed regime, etc., have impacted on sustainability of production systems.

4. Farmers have outpaced extensionists and become technology driven in the most favoured area. However, they have not been able to keep abreast of dynamic changes in their technical environment. The information processing abilities and technical skills required for coping with sustainability issues are complex and need a qualitatively improved research and extension effort.

5. Complex diversities and risk proneness in the rainfed region give rise to a myriad of interrelated problems that the professionals have not understood precisely. Indigenous knowledge systems, farmers' coping strategies need to be integrated with formal R&D by approaches which have systems perspective and participatory orientation.

6. A few selected pockets of high value commercialized agriculture have emerged with market sensitive production systems. The organized sector entering high value agriculture have a back-up

support of technology from advanced countries. Managerial competence, technical skills and search capacity of large individual producers is well developed. They are not depending on public systems. The unorganized groups obtain their hardware technologies from private sector. The role of public system is limited to conducting strategic and anticipatory research.

7. The Indian National Agricultural Research System is institutionally mature. It has a hierarchy of institutions capable of undertaking basic, strategic, applied and adaptive research. A synergic relationship between the institutions will assist in avoiding duplication and overlaps. The Zonal Research Station (ZRS) is emerging as a pace-setter for location specific and need based technology development. All future efforts of integrating research and extension for agricultural technology development should originate at the ZRS. However, the ZRS researchers need methodological reorientation, adequate untied funds and mobility to initiate farmer-relevant technology development research.

8. While the T&V System continues to grapple with internal issues that have arisen due to the extension reform, the States have been responding to several external compulsions too. Consequently, the organizational structure and extension responsibilities have undergone modifications. SMS continues to be the weakest link. General disillusionment with the concept of "contact" farmer has motivated a few States to experiment with group based approaches.

9. Initial enthusiasm to observe the rigid and uniform schedules of training and visits is waning. Lack of career advancement opportunities and absence of performance related incentives and rewards have further demotivated the Agricultural Extension Officers (AEOs) and Village Extension Workers (VEWs) — the cutting edge of the public extension system.

10. Non-Governmental Organizations (NGOs) have increasingly adopted "bottom-up" approach of agricultural technology development and dissemination in tandem with empowering local community. They have been particularly innovative in participatory methods and farming systems approach, dissemination methods, improved training methods and combining human resources development with technology transfer. A few innovative NGOs are in a position to collaborate with the public extension and research systems.

11. The public sector cannot be expected to attend to the

technological and extension needs of all farmers and farmer categories. There will be an increased participation of the private sector in extension related activities. NGOs, on the other hand, either in tandem with Government services or by themselves are increasingly providing extension services in the difficult rainfed regions. The public sector should evolve new roles in policy development, coordination and regulation of extension services.

12. The public sector extension should embrace, besides technology transfer, other roles like human resources development, broad basing and farming systems perspective and gender-differentiated strategies.

13. Client participation in programme formulation in a sustainable manner can be achieved through farmer groups rather than through individuals. The poor and disadvantaged population cannot bear the cost of organizing. There are other well known problems of group formation which can be overcome if extensionists link with NGOs. Groups formed around joint management of a natural resource, around technologies requiring joint management around an economic enterprise, or around technological options, suffer to a lesser extent with group formation problems.

14. Farmers organizations can be important in creating demand on research and extension system and be a vehicle for developing technical and managerial skills of members. However, extension and research workers will require flexibility and new skills if they are to work successfully with a wide array of special groups.

15. Farm extension volunteers selected by, partly compensated by, and accountable to local farmers organizations can increase the relevance and reduce the burden on public budget, while increasing local self-reliance.

16. Farmer participation in agricultural technology development and dissemination should begin from diagnosis to planning and designing technological solutions, to implementation, to evaluation and feed-back into dissemination. However, participation process should also empower the local groups to exert demand-pull on research and extension system. But the success of farmer participatory process will depend on striking a balance between science-led and bottom-up approaches of agricultural technology development.

17. Innovative NGOs are well placed in providing a range of skills that the public sector currently lack, such as group formation,

HRD, experiential learning, participatory approaches, sustainable livelihoods.

18. Decentralization of programme planning in public extension and research allows priorities to vary as per the local needs. The farmers organizations will play a vital role in extension programming and formulating VEWs agenda. However, the success of decentralized extension ,programme planning hinges on adequate technical and financial support to the VEW. Appropriate performance-linked incentives and reward system should be instituted to reinforce extensionists farmer-participatory behaviour.

19. Farmers organizations should be involved in monitoring and evaluation of the extension programmes. Monitoring of impacts should use indicators like "activities initiated in response to local level needs", "resource mobilization for problem solving", "impact per person", "impact per unit area" or "effectiveness in solving some of the prioritized problems". Horizontal diffusion, farmers adaptation behaviour should be concurrently evaluated for mid-course corrections.

20. Each Department of Agriculture (DOA) should earmark 5 per cent of its annual plan budget for flexible funding of innovative ideas that emerge from the field level. All central and State sponsored special programmes should set aside 10 per cent of their current outlays for supporting farmer participatory approaches.

21. In order to create an enabling environment for State departments, NGOs and Agricultural Research Institutions to reorient themselves to farmers' problems and become client-accountable, it is recommended that the Government of India should launch a National Programme on Farmer Participatory Extension and Research.

22. The State research agricultural resources should be allocated exclusively for farmer participatory research and adaptive research at the SAUs' Zonal Research Stations. The State should fund research on contractual basis.

23. A Joint Extension-Research Management Board should be constituted to oversee, monitor, mobilise and allocate resources for research and extension at the zonal level. Research management capability should be strengthened to monitor progress of research programmes and ensure their farmer-relevance.

24. A clearing house mechanism at the national level for preparation of reader--friendly documents on innovative extension

strategies, approaches and methods and extension management for extension staff at different levels should be set up. Establishment of extension communication unit for transformation of technical information to extension messages and dissemination of technology message, extension message and marketing information is an essential component of an effective farmer participatory agricultural extension system.

25. A substantial training effort is needed to upgrade the technical, managerial and organizational skills, knowledge and qualifications of field and junior extension personnel.

26. Experimentation on cost reducing extension strategies should be designed and implemented through the farmers organizations. Farm extension volunteer concept should be implemented.

27. Continuation of reform processes to make the public extension and research systems farmer participatory and client accountable should receive strong policy support.

28. In irrigated situation, researchers should design a farmer monitoring of sustainability issues and thereafter launch anticipatory research on sustainable agriculture. The best way of improving professional competence is to encourage extensionists team at block level to collaborate with SAU scientists in joint professional activities like problem diagnosis, prioritization and planning exercises, formal collaboration in on-farm research and dissemination activities, joint training and seminars etc.

29. In rainfed situations, researchers should blend information research on indigenous knowledge systems with formal research. Increasingly, a farming systems perspective should be adopted by both extensionists and researchers and there should be joint professional activities. VEWs should be enabled to acquire multiskills. In districts with non-crop dominated enterprise, lead department function should be handled by the dominant enterprise department.

30. In commercial agriculture regions, public systems have a very limited role. The private sector should be encouraged to assume important new research and extension activities.

1

INTRODUCTION

India began institutional reforms to strengthen its agricultural extension services through two pilot projects in the States of Rajasthan and Madhya Pradesh in 1974. The new approach, now famous as the Training and Visit (T&V) System, introduced a series of changes that were organizational and operational in nature. Restructuring the State Departments of Agriculture (DOA) into a single line of command with exclusive focus on extension was the organizational reform. The new operational features introduced were : a fixed schedule of properly supervised visits, reasonable staff-farmer ratio, minimal paper work and administration, explicit field orientation and mobility. In addition, the T&V system sought to strengthen extension-research linkage through monthly workshops of extension and research staff, joint field visits and formation of diagnostic teams. This reform has since been implemented in 15 other States. These initiatives were supported through the World Bank aided National Agricultural Extension Projects (NAEP).

1.02 In 1979, India embarked on yet another ambitious project National Agricultural Research Project (NARP)—designed to strengthen the capability of State Agricultural Universities (SAU) to conduct decentralized research to support the reformed State-wide agricultural extension services. The two phases of NARP were supported by the World Bank. NARP aimed at strengthening the office of the Directors of Research at the main SAU campuses and to establish, staff and equip Zonal Research Stations (ZRS). Support to the T&V system of extension service was conceived of as decentralized, farmer-oriented adaptive research based on the distinct needs of the agro-ecological zones. Special emphasis was placed on foodgrains, pulses and oilseeds, especially those grown under rainfed conditions; attention was also given to the integration of cropping patterns and animal husbandry.

1.03 Notwithstanding these institutional reforms there is widespread concern among policy-makers, agricultural administrators and researchers about the slackening of the pace of technology adoption by farmers. This draws our attention to the relevance of the technology currently on offer and the effectiveness with which technologies pass from research to extension.

1.04 The impetus for continued stress on reform of public agricultural extension and research stems from the challenges of Indian agricultural development in the present decade and beyond. For instance, in the irrigated regions, the productivity has hit a plateau while resources have depleted and their quality is also fast deteriorating. Moreover, as the physical limits to India's irrigation potential are achieved and as the lack of recent major breakthroughs in the genetic potential of the major field crops suggests, two-fold strategies are required to be pursued in the future; first, concerted research efforts to retain the productivity advantage in the irrigated plains; and secondly, launching a major research and extension initiative on rainfed areas for enhanced agricultural production.

1.05. The "second generation" problems of soil and water in the most favoured areas calls for science led but farmer-participatory experimentation. The professional coded knowledge systems on rainfed areas, on the other hand, do not match the richness of indigenous knowledge about environmental diversities and farmers' coping strategies. A resource-based farmer-participatory research and extension strategy that blends informal and formal R&D is most likely to make the difference in the difficult region. Thus, newer challenges confronting India's agriculture warrants greater responsiveness from its agricultural extension and research systems.

1.06. There are considerable opportunities to improve quality, management and cost effectiveness of agricultural extension and research. A major challenge is to incorporate more effectively farmers' perceptions about their problems into the programming of public sector extension and research services, thereby making the systems increasingly farmer-driven and controlled. In operational terms, this would mean developing procedures that encourage farmer participation in planning and management of agricultural extension and research so as to reflect diverse needs arising due to interactions between biophysical and socio-economic contexts surrounding farmers' decision-making situations.

1.07 Farmer participatory approaches have been found to

improve cost effectiveness of responsiveness. They draw antecedence from social sciences and farm management economics. "They are motivated by a concern that conventional quantitative and neutral research methods tend to preserve social inequality" (Farrington and Martin, 1988). Farming systems research extension (FSRE) has evolved out of conventional agricultural research systems and is seen as "an approach to agricultural research and development that views the whole farm as a system and focusses on (i) the inter-dependencies between the components under the control of members of the household, and (ii) how these components interact with the physical, biological and socio-economic factors not under the household's control" (Shaner *et al.*, 1982).

1.08 A large number of proponents of FSR or FPR strongly believe that alternative or improved technologies can be developed only by involving farmers in the whole process of agricultural technology development and dissemination. This requires major changes in attitudes, approach and role of researchers and extensionists (Chambers and Ghildayal, 1985; Chambers, Pacey and Thrupp, 1989) and invites farmers to set the research/extension agenda. Development of this approach has focussed strongly on participation of farmers and inclusion of indigenous knowledge and farmer experimentation in the research process (Haverkort *et al.*, 1991). Over the past decade some progress has been made in the incorporation of FSR and on-farm trials into National Agricultural Research System (NARS) and linking farmers, extensionists and researchers. The International Service for National Agricultural Research (ISNAR) studies found that the increased farmer participation in research and technology transfer has several advantages (Merrill Sands and McAllister, 1988 and Merrill-Sands and Kaimowitz, 1989).

1.09 In India farmer participatory research in its narrow sense can be traced back to experiments on cultivators' field. However, systematic work began in involving farmers only in the mid-1980s with donor-supported farming systems research/extension projects in a network of State agricultural universities (SAUs) in eastern India (John 1991, 1996). The incorporation of farmers into research planning and priority setting process has proved to be very difficult, and the major decisions on the strategy direction of the programme remain in the hands of researchers (Sara, John, Kumar, 1992).

1.10 However, a few developments in India have been

encouraging and exciting. NGOs have innovated participatory methodologies and have experimented with farmer groups as an appropriate unit of analysis of collaboration. This has been an important step forward as the NGOs have sought to empower more relevant and homogeneous groups of farmers to articulate their needs. The role of NGOs as effective agents for relevant, client oriented technology generation has been reinforced by cross country studies initiated by the Agricultural Administration (Research and Extension), Network, London. Several NGOs have entered into collaboration with public systems too.

1.11 International centres like IRRI and CIMMYT have recently joined hands with Indian researchers to investigate sustainability of rice-wheat systems in irrigated plains. Farmer participatory diagnostic studies in Haryana, Western Uttar Pradesh, Tarai Region and Eastern Uttar Pradesh have revealed that besides short-run production problems, it is the farmers practices of nutrient mining, unregulated groundwater extraction, continued cereal-based cropping patterns which have not only depleted the resources but also deteriorated the quality of resources in the Green Revolution Regions.

1.12 The issues emerging from Indian experience of farmer participatory technology development and dissemination fall into three major areas.

1.13 The *first* issue concerns reorganizing institutional processes to accommodate flexibility in approaching emerging problems of both well-endowed and not so well-endowed regions. This applies both to research institutions and extension services. Besides, they also need to develop formal linkages with other change agencies in the rural setting. Institutional values and reward systems need to be reviewed to encourage farmer participation, systems thinking, multi-disciplinary teamwork and outputs.

1.14 The *second* relates to the approach and methods of research/ extension. The Indian experience of SAUs and NGOs points to the need to be concerned with functional participation vis-a-vis expowerment of farmer organizations. This means extensionists and researchers may have to enlarge their role beyond technical change.

The *third* addresses the present and future needs of farming communities, both in production, and consumption patterns and the need to manage quality of resources and the environment in a sustainable manner. At the present juncture, it is no longer possible to isolate the perceived needs of individuals and communities from the impact they have on non-renewable resources.

1.15 The scale of reforms in India brought through NAEP and NARP is impressive. A strong systematic and sustained working relationship has been constituted between agricultural extension and research services in most States. Although the inter-State quality of both extension and research operations varies, there is a general appreciation of the institutional developments achieved during the past 25 years or so. There is a strong foundation and a decentralized structure available in the field which the present study reviewed in order to suggest improvements in its functioning. More specifically, the present study had the following objectives:

(i) To generate alternate options that encourage farmer participation in planning and management of agricultural extension and research:
(ii) To suggest institutional mechanisms to make research scientists and extension professionals accountable to the client community; and
(iii) To design pilot projects for empirically evaluating feasibility of different options of farmer-driven and farmer-accountable agricultural extension and research.

1.16 This report contains the findings of a review carried out by a Team of 5 consultants (Annexure II) appointed by the Government of India. The study team was asked to examine the existing structures of agricultural extension and research and their operations in different agro-ecological zones, to analyze the present situation and to propose options for improving farmer participation both in research and extension. The following chapter describes the methodology that was adopted to conduct the study. Chapter 3 explains the rationale of creating generic agricultural situations and contains a diagnosis of emergent problems in the different situations which demand a major reorganization of research in irrigated and assured rainfall region, rainfed agricultural region and commercial agricultural niches. Chapter 4 deals with an institutional analysis of public agricultural research and extension systems and non-governmental organizations active in participatory agricultural technology development and draws out issues for design of farmer participatory processes. Chapter 5 attempts to weave together a conceptual framework that is grounded in the analyses of the preceding two chapters, i.e., agro-ecological and institutional analyses, and sketches strategic options that are most likely to make public extension and research systems farmer participatory and client accountable. In this chapter agricultural situation specific options are also evaluated. In the final chapter pilot project outlines are developed for enabling experimentation with farmer participatory processes.

2

METHODOLOGY

2.01 The study team had access to a desk review of literature prepared by MANAGE, Hyderabad (Annexure III). It contained a review of literature documenting experience in farmer participation and control of public sector agricultural extension and research and Non-Governmental Organizations (NGOs) initiatives.

2.02 The study team also examined the following review of Indian Agricultural Extension and Research :

(i) Feder, G. and Slade, R., 1989, "Institutional Reform in India: The Case of Agricultural Extension", The World Bank, Washington, D. C. mimeo.

(ii) GOI, 1988. "Report of the Screening Committee on the Findings of the World Bank Mission's Review of the Agricultural Extension Sector in India." Ministry of Agriculture, Department of Agriculture and Cooperation (Extension Division), New Delhi.
(Also known as K. N. Ardhanareswaran Committee Report)

(iii) ICAR, 1988. "Report of the ICAR Review Committee." ICAR, New Delhi.
(Also known as G. V. K. Rao Committee Report)

(iv) ICAR. 1989. "National Agricultural Research Project-II: Midterm Review, Project Unit (NARP), ICAR, New Delhi.

(v) Randhawa, N. S. 1987. "Agricultural Research in India: An Overview of its Organizations, Management and Operations," Research and Technology Paper No. 3, FAO, Rome.

(vi) World Bank, 1988, India: Agricultural Extension Sector Review, Washington D. C., mimeo.

2.03 The study team identified a critical set of issues, based on the desk review, that were germane to the farmer participatory study (Annexure-IV).

2.04 The issues formed the basis of a two-day National Workshop with policy-makers, agricultural administrators, agricultural scientists and extensionists, and representatives from NGOs (Annexure-V). The brainstorm, with a specially invited expert panel, helped the study team to identify farmer participatory projects being implemented in India. In addition, the discussions proved useful in formulation of a checklist of issues for probing relevant dimensions of farmer participatory projects (Annexure-VI).

2.05 The discussions also revealed that currently India is experimenting with a veritable mix of institutional arrangements to enhance the responsiveness of agricultural technology generation and dissemination systems to varying farmers needs. These initiatives were classified into the following categories (Annexure VI).

(i) Public Sector Initiatives;
(ii) Cooperative Sector Initiatives;
(iii) Growers' Association Initiatives;
(iv) Private Sector Initiatives; and
(v) NGO Initiatives.

2.06 The study team was divided into two groups, which in turn visited a few selected farmer participatory projects in Andhra Pradesh, Karnataka and Gujarat, being implemented in all sectors except the public sector. The approach has been to learn from key officials' experiences in implementing farmer participatory projects and examine feasibility of their replicability elsewhere. The major focus of the field visit was:

(i) To understand how farmer participatory approaches were initiated and integrated into organizational (Cooperative/ Growers Association/Private/NGO) activities;
(ii) To identify problems encountered in implementing farmer participatory approaches in technology generation and dissemination, and to diagnose institutional factors that led to emergence of such problems;
(iii) To discern institutional/individual or group responses to overcome barriers to client participation; and
(iv) To distil lessons for successful management of linkage

mechanisms: Research-Extension, NGO-GO, Cooperative-Public research and extension, Private-GO/Private-NGO research and extension.

2.07 The study team visited the following agencies to learn from their farmer participatory experiences:

Cooperatives	Oilseeds Growers' Federation, Ahmedabad and Mehsana, Gujarat.
Private Sector	KRIBHCO - Sponsored Rainfed Farming Project, Dahod, Gujarat.
	ITC Agro Business Divisions' Sunflower Seed Project, Kurnool, Andhra Pradesh.
NGO-GO Collaboration	MYRADA - Dryland Development Board Project on Watershed Management, Kamalapur, Gulbarga, Karnataka.
	Sri Aurobindo Krishi Vigyan Kendra, Gaddipalli, Nalgonda, Andhra Pradesh.
NGO-Public Research/ Extension Collaboration	Action for World Solidarity/Deccan Development Society led consortium project on Integrated Pest Management of Red Hairy Caterpillar of castor in Warangal, Andhra Pradesh.
NGO	Aga Khan Rural Support Programme, Ahmedabad and Bahruch, Gujarat.
	Sadguru Water Development Foundation, Dahod, Gujarat.
	Bharatiya Agro Industry Foundation,. Ahmedabad/Baroda, Gujarat.
Growers' Association	Agri-Horticultural Society and FRESH Marketing Society, Hyderabad.

2.08 Analysis of the most promising case experiences of farmer participation (Annexure-VI) and approaches identified through the desk review helped the study team to identify major issues (Annexure-VII) for exploration during the field work. Three major constituencies of the public agricultural extension and research services, viz., farming communities, researchers and extension professionals from distinct agro-ecological situations were met separately.

2.09 The suggestions from the earlier National Workshop were also instrumental in delineation and selection of various agricultural situations. The agricultural situations visited were :

(i) *Assured irrigated areas* : (Command areas in Tanjavur, Tamilnadu, Tarai regions of Uttar Pradesh, tube well irrigated plains of Western Uttar Pradesh and shallow tube well irrigated alluvial plains of West Bengal).

(ii) *Assured rainfall region* : (Vidarbha region of Maharashtra, Brahmaputra valley in Assam and Coastal areas of West Bengal and Tamilnadu)

(iii) *Hilly Regions* : (Meghalaya, Kumaon and Garhwal ranges in Uttar Pradesh and the Nilgiri hills in Tamilnadu).

(iv) *Rainfed region* : (Drought prone regions of West Bengal and Tamilnadu), and

(v) *Arid region* : (Deserts of Rajasthan).

2.10 Participatory Rural Appraisal (PRA) techniques were employed to assess farmers analysis of their environment as well as their perceptions on current and emerging agricultural problems. Participatory Rural Appraisal also enabled the study team to obtain farmers' perceptions on efficacy of existing institutional support and feasibility of their meaningful participation in agricultural extension and research. Interactive workshops with groups of researchers and extension professionals assisted the study team to understand the functioning of agricultural extension and research structures as well as their strengths and weaknesses. Lessons distilled from farmer participatory projects were shared during the workshop. Several options of making the existing agricultural extension and research services farmer participatory and client driven were also explored. The team also visited public agricultural extension and research systems initiated experimentation on farmer participatory approaches.

2.11 Emergent analysis and tentative formulations of recommendations were shared, at the second National Workshop, with eminent scientists, extension professionals, agricultural administrators and a few representatives of NGOs and private sector. The major concern during the workshop was to critically examine the implementability of recommendations and the feasibility of alternate options to make public agricultural extension and research services farmer-participatory and client-accountable in the context of complex political and bureaucratic realities (Annexure-VIII).

2.12 In the present report the study team has attempted to moderate its recommendations in the wake of critical suggestions received during the workshop.

2.13 MANAGE circulated the final report widely among relevant officials from Government of India, State Governments, State Agricultural Universities, Indian Council of Agricultural Research and NGOs. This was followed by a National Conference convened by the Government of India in June 1993 where the report was approved. Action is being taken at Government of India level for implementation of the recommendations.

3

DIAGNOSIS

Concepts of Zonation for Agricultural Extension and Research

In India several approaches have been used in classifying agro-climatic resources into different zones. The Agro-climatic Regional Planning Unit (ARPU) of the Planning Commission identified 15 resource development regions in the country based on agro-climatic (soil, topography, climatic and water resources) and technological considerations, rather than on social and administrative criteria. The main purpose of the classification scheme was deconcentration and regionalisation of development planning through Regional Agricultural Planning Teams.

3.02 The Indian Council of Agricultural Research under the aegis of NARP classified the country into 126 macro agroclimatic zones. The classification was based on identification of contiguous areas within the administrative boundaries of a State—homologues of topography, rainfall, soils, cropping patterns and irrigation availability.

3.03 The National Bureau of Soil Survey and Land Use Planning (NBSSLUP) of the ICAR has grouped India into 21 macro ecological zones primarily on edaphic factors but superimposing other criteria such as, growing period as a principle component proxying for effective moisture regime and physiography. The delineated boundaries were adjusted to correspond with the administrative district boundaries.

3.04 Recently, the Planning Commission and ICAR reached a consensus on the NBSSLUP ecological classification of 21 macro zones.

3.05 The study team initially used the ICAR's macro-level classification to identify the regions to be visited for reviewing the existing institutional arrangements of agricultural extension and research and the extension-research linkages. However, analysis of agricultural situations and differential agricultural extension and research needs of various regions led the team to classify the country into following generic groups (Table 3.1):

(i) Assured rainfall and irrigated areas;
(ii) Rainfed areas;
  (a) Rainfed crop-based farming systems,
  (b) Pastoral systems,
  (c) Rainfed horticulture based farming systems of hills,
  (d) Rainfed aquaculture-based farming systems of the coastal regions, and
(iii) High value commercial agriculture regions.

3.06 While the concept of ZRS is strategic, both as a pace-setter for location specific and need based technology development as well as creating ripple effect through effective extension-research linkages, our generic classification draws attention to differential research and extension needs in two ways. First, meso-level ecological classification does not capture the heterogeneity, though attempts have been made to delineate "farming situations" in a static manner. Second, it acknowledges that not only research programme planning and extension strategies ought to be responsive to technological development and dissemination needs that vary greatly within a zone, but also that the existing structures require loosening through design of flexible operating procedures.

Agricultural Situation Analysis

3.07 The farming situations (contained in the Status Report) have been delineated on the basis of agro-ecological situations influencing the cropping patterns in different meso-level zones. Edaphic factors like soil structure, texture, soil depth, soil reaction, drainage, landscape and variations in moisture regime associated with rainfall and irrigation were used flexibly in the determination of a farming situation. Our agricultural situations, however, cut across the meso zones and are generic and broader in concept than the farming situations. Asssured moisture regime, complexity of the farming systems reflecting the magnitude of ecologically-induced risk and

TABLE 3.1 : **Classification of Agricultural Situations in India**

Distinctive Features	Assured Irrigation System	Rainfed Farming System	High Value Commercial Crop System
1. Literacy	High	Low	High
2. Economic Status	Affluent	Poor	Affluent
3. Production System	Simple Monocropped	Complex Diverse Enterprise	Specialised Commercial Enterprise
4. Dominant Occupation	Crop Husbandry	Crop Husbandry and Livestock/Horticulture Aquaculture	Crop/Tree Husbandry
5. Risk Factor	Low	High Drought/Flood Prone	Low to Medium
6. Ecological Fragility	Low	High	Low to Medium
7. Topography	Usually Flat Lands	Usually Undulating Lands	Usually Flat Occasionally Hilly
8. Soil Status	High Fertility	Problem Soils	Fertile
– Special Problems	Waterlogging Salinity	Acidic Soils	
– Erodability	Low	High	Low to Medium
9. Input Consumption	High	Low	High
10. Productivity	High mean and Low variation	Low mean and High variation	High mean and Low variation
11. Role of Indigenous knowledge Systems	Obsolete	High	Low

(Contd.)

TABLE 3.1 *(Contd.)*

Distinctive Features	Assured Irrigation System	Rainfed Farming System	High Value Commercial Crop System
12. Social conditions	(a) High evolved farmers' pressure group (b) Highly articulate individual farmers	(a) Low capacity among farmer to articulate requirements (b) Weak farmers organizations, NGOs, help to speak on behalf of the poor and for the long term strengthening (a) & (b)	Highly evolved farmers organizations; Highly articulate indi vidual farmers.
Illustrative List of States/Regions	– Punjab – Haryana – Western U.P. – Krishna-Godavari belt of A.P. – Thanjavur belt of Tamil Nadu – Central Bihar	Rajasthan Gujarat M.P. Eastern U.P. Bihar Bengal Karnataka	Himachal Pradesh Parts of Punjab Parts of Maharashtra Parts of A.P. Parts of Tamil Nadu Kerala

farmers' risk-coping mechanism, and market-orientations of the farming community are the three major determinants of the generic classification. The classification schema also captures variations in literacy level, quality of life as well as items of infrastructure development.

Assured Rainfall and Irrigated Areas

3.08 Adequate moisture regime facilitates farmers to raise either a sequence of two crops or three crops in the assured rainfall and irrigated areas. The Indo-Gangetic alluvial plains, the coastal inland and the peninsular plateau regions are the most favourable areas where the impact of green revolution technology have been significant. While the alluvial plains receive a uni-modal precipitation, in the range of 800-1500 mm, the latter two have a bi-modal monsoon (3000-4000 mm), i.e., around 60-70 percent of the rainfall is received during the south-west monsoon and the remainder during the north-east monsoon. In addition, these areas are interspersed with both major and minor irrigation facilities, such as canal, tube wells, tanks and dug wells.

3.09 Supplemental irrigation and favourable residual moisture enable farmers in the asssured rainfall areas to raise successfully a two-crop sequence. Whereas, in irrigated areas, a multiple cropping pattern pivoted around miracle seeds of cereals is popular.

3.10 Two of the major sources of agricultural growth in the most favoured areas – the spread of modern varieties and rapid increase in fertilizer consumption – have already beeen substantially exploited (Table 3.2). Genetic gains in yield potential in successive generations of improved varieties have slowed an increasing proportion of research resource is now being devoted to "maintenance research" to protect yield gains against breakdown of resistance to pests and diseases.

3.11 Though major sources of growth, viz., genetic gains, fertilizer responsivenesss and assured moisture regime have been exploited to a large extent, substantial opportunities exist to increase production through increased yields, reduced costs through efficient use of purchased inputs and water, and selection of efficient and improved cropping systems. The extension strategies in this context need synchronizing technology advise with market intelligence.

3.12 A wide array of second generation problems have begun to

TABLE 3.2: **Productivity Levels in Most Favoured Areas**

Crop	State	Yield kg/ha	All India Average kg/ha
Rice	Tamilnadu	3256	1756
Wheat	Punjab	3621	2117
Maize	Andhra Pradesh	2643	1606
Sorghum	Tamilnadu	2222	864
Rapeseed Mustard	Gujarat	1165	826
Groundnut	Maharashtra	1673	1484
Sugarcane	Tamilnadu	101228	65375
Cotton-Lint	Punjab	570	265

emerge in these areas that impact on sustainability of production systems. Increased yields and cropping intensity coupled with farmers' practices of imbalanced nutrient management–high dose of chemical fertilizers with low or no application of organic manure – has tended to deplete soil of both macro- and micro-nutrient reserves, i.e., nutrient mining.

3.13 Same crop sequence – paddy-paddy-paddy or paddy-wheat— and control of broad leaf weeds has reduced the competition for grassy weeds. As a result incidence of specific weeds have increased. The changing weed-regime and their increasing population demands an integrated understanding and weed management strategies.

3.14 The continued practice of same crop rotation over the years also creates a conducive environment for pest build up in the soil. Indiscriminate use of synthetic pesticides in paddy, cotton, oilseeds and vegetables, in turn, has not only induced emergence of newer biotypes and strains that are resistant to recommended chemicals, but has also resulted in destruction of natural predators and parasites.

3.15 In major irrigation command areas, insufficient attention to drainage system and absence of scientific and equitable water distribution and management practices have given rise to problems of protracted waterlogging and increased incidence of soil salinity. Various estimates suggest that about 25 percent area under major irrigation command area have been turned into wastelands.

3.16 Increased supplies of tubewell water, flat rate of water pricing and power tariff and higher cropping intensities complicate water management in irrigated areas. Seasonal and permanent draw down of acquifers in different irrigated areas raises serious questions about sustaining the level of productivity and production in the grain baskets of India.

3.17 The research system in such areas besides killing repetitive research projects, may design farmer participatory monitoring systems that allows it to initiate anticipatory research of strategic nature.

3.18 Farmers in assured rainfall and irrigated areas have bypassed the conventional extension system and established productive linkages with the research systems. Even smaller farmers have become science-driven and are better informed than the VEOs. Besides, they have access to well developed infrastructural facilities like, roads, markets for inputs and outputs, credit institutions, power supply, etc.

3.19 Farmers, however, have not been able to keep abreast of dynamic changes in their technical environment. Productivity in the most favoured region has either hit a plateau or declined over time due to depletion of soil nutrients, pathogen build-up, draw down and changing weed-regime. Research on sustainable agriculture, therefore, should be accorded high priority.

3.20 The second generation inputs often require more information and skills for successful adoption than the initial round of introduction of miracle seeds, which were well known for their wider adaptability and nitrogenous fertilizer. The use of more complex fertilizers, micro-nutrients and pesticides often interacts strongly with variation in micro-climate and pest incidence. Hence, individual farmers need to adapt the technology to their own requirements. Moreover, all second generation problems demonstrate common pool resources characteristics requiring collective action. Mobilizing of collective action, however, will depend on heightened awareness among farmers about their interdependence. In the absence of National Policies to ensure efficient input-use, rather than designing external enforcement mechanisms, interdependent individuals, if facilitated, can institute self-binding norms and are capable of self governance of local resource management institutions.

3.21 Farmers' technical skills, information processing ability and management competence are of higher order than the first round of diffusion of innovation strategy. The value of traditional knowledge systems and intimate knowledge of their environment in the science-based agriculture tends to depreciate rapidly as new problems emerge and subsequently new inputs and cropping patterns are introduced.

3.22 The Transfer of Technology (TOT) mode of experiment station research used by agricultural scientists and the reliance on technocratic progressive 'contact' farmer approach used by extension service appear inadequate models to guide us in addressing the sustainability issues.

3.23 The VEO – the lynchpin of the existing technology development and dissemination system – is ill-equipped either to monitor the dynamics of changes in the technical environment or to provide credible farm advisory services.

Rainfed Areas

3.24 In India almost 70 per cent of the arable land is dependent on the proverbial vagaries of monsoon. Rainfed agriculture accounts for over 40 per cent of the total production of foodgrains. However, the regions representing rainfed areas have been bypassed by the green revolution technologies.

3.25 One major reason for experimental station-based technologies not making a dent is the inability of researchers to stimulate diverse bio-physical conditions prevalent in the rainfed areas at the experiment station. Unlike in most favourable areas, where the environment is homogeneous over a vast area and similar to the conditions existing at the experiment station, the rainfed situation is characterized by heterogeneous micro environments. Farming community rely on agriculture practised under, what has now come to be characterized as the complex, diverse and risk-prone (CDR) conditions for their livelihood. A wide range of enterprise-mix prevails, from pastoral-based systems in arid Western Rajasthan, to the horticulture-based systems in Himachal Pradesh and the North-East, to the aquaculture-based system in the coastal areas of West Bengal, Orissa and the peninsular India, to the crop-based systems in larger tract of the rainfed semi-arid and humid tropic. The farming community has also to cope with a combination of moisture stress, problem soils, undulating topography, erodibility of soils and high water run-off in drought-prone regions. On the other hand, in the flood-prone basins the problems are: abundance of water due to drainage congestion and waterlogging and unpredictable rainfall that may induce drought and flood conditions in a single production cycle. Obviously, agricultural extension and research should take cognizance of these diversities and adopt a resource-based strategy of technology development and dissemination. Instead of modifying farming environment, which is very difficult in the CDR agriculture, technologies should be modified to fit the existing environment.

Rainfed Crop-based Farming Systems

In a large proportion of rainfed area, the farming communities eke out their livelihood predominantly through monocropping. The semi-arid regions of Eastern Rajasthan, Gujarat, Andhra Pradesh, Karnataka, Tamil Nadu, Maharashtra and Madhya Pradesh receive a uni-modal rainfall, that ranges between 500-1000 mm; the upland plateau spread over vast tribal areas in Madhya Pradesh, South Bihar, Purulia in West Bengal and Orissa receive annual precipitation between 1000-1200 mm. The rainfall exceeds 1200 mm in the humid tropic regions of Uttar Pradesh, Bihar, West Bengal, Assam and Orissa. Around 70 to 80 per cent of the annual precipitation is concentrated in 2-3 months, which is usually lost as run-off. Even this rainfall is erratic and uneven. For example, in Rajasthan, 3 years in a cycle of five years are invariably drought years, when the actual precipitation is lower than the average. In Karnataka, it is 3 years in 7 years, while in South Bihar, it is 5 years in 8 years. Thus, the farmers face a cycle of good, normal, bad and very bad years in terms of crop production due to aberrations in the weather.

Farmers' risk coping mechanisms include inter- and mixed-cropping strategy with a low but stable productivity at subsistence level. Seasonal migration of male members during bad years makes a large number of vulnerable households women-headed. Besides reproduction and nurturing responsibilities, women in this region play an important productive role also.

Pastoral Based Systems

Pastoral systems primarily exist in arid desert of Western Rajasthan and North Gujarat. It receives scanty uni-modal precipitation that is well below 500 mm annually. The pastoral community experience frequent food deficits, were among the hardest hit during the successive drought years in the 1980s and their production systems suffer serious short, as well as long-term perturbations. Consequently, the community is regularly on the move in search of greener pastures.

On-going environmental degradation and fragility of the ecosystem calls for a reorientation of extension towards environmental protection and enhanced resource conservation activities with the potential of augmenting the local livelihoods strategies.

The indigenous knowledge of the pastoral community on animal breeding, animal nutrition and husbandry and local coarse grain and secondary crops with high fodder value is yet to be integrated into the formal agricultural research and extension strategy.

Horticulture-based Farming System of the Hills

Specificities of the hilly horticulture-based farming system arise due to interactions among features like inaccessibility, fragility, marginality, heterogeneity, micro-niches and human adaptation mechanisms. Their implications for research and extension remain to be understood in a holistic sense. Sectoral extension programmes of soil conservation, reforestation, vegetable and tree-crop production and agricultural production have short-term focus, ad hoc basis, are resource extractive in nature and replicative of designs and experiences of plains that are often untested and unsuited to the hilly terrain. Both for pastoral and hilly horticultural systems, a holistic natural resource management perspective rather than crop-based production systems is likely to yield rich dividends.

Aquaculture Based Farming Systems

Coastal areas (West Bengal and Orissa) with recurring incidence of incursion of saline water pose yet another set of challenges to resource-poor farmers in eking out livelihood. Small fragmented holdings, high population density, high pressure to raise paddy, preponderance of rice pests and diseases, protracted water logging, unsuitable water for irrigation and saline affected patches compel the farmers to rely on mono-cropping with traditional pisciculture. While high-productivity potential aquacultural technologies have been generated under controlled conditions of experiment stations, they have not been integrated to farming situations prevalent locally.

3.26 Commonality across the four distinct subsystems are: relatively poorly developed infrastructure facilities, low literacy levels, larger proportion of marginal and small farmers, greater incidence of fragmented land holding, lower population densities, subdued and interlocked markets and absence of farmers' organizations. Thus, not only farmers are not able to articulate their needs, but institutional capacity to monitor and respond to latent needs are conspicuous by their absence.

3.27 Often, researchers' problem definition is incongruent with farmers' perception of problems. In the rainfed situation, as a result, professional led research and extension is insensitive to farmers' perceptions and is unable to develop inter-disciplinary perspective in adapting technology and exploring local environmental conditions. The underlying principle can be traced to observance of uniform policies and methods of research and extension, which were developed for most favoured irrigated areas but inappropriate for CDR agriculture (Gupta, 1987).

3.28 Increasingly, it is becoming obvious that mainstream agricultural research and extension strategies associated with the Green Revolution are not appropriate to more difficult and riskprone rainfed areas. The unfavourable, marginal and complex biophysical environments in rainfed areas, multiple objectives of farmers that is embedded in their survival strategies and complex interactions among the enterprise-mix cannot be simulated on the experiment station. Nor will the current obsession with discrete component research with yield as sole evaluation criteria assist farmers in improving their existing production systems. As a result, extension strategies that promote package of technologies not well matched to the inherent complexities of local farming systems are likely to be rejected.

3.29 Of late, a consensus seems to have been reached that livelihoods under rainfed regions can be enhanced through approaches which are *systems-oriented and participatory*. The usefulness of farmers' knowledge, the partnership between farming community and research/extension professionals, emphasis on identifying farmers' problems as a starting point for participatory technology development, a broader rather than a "model" approach to research and extension are some of the salient characteristics of the emerging approaches to research and extension.

3.30 Simultaneously, the extension strategy would also need to enlarge its narrow focus on crop to include livestock/horticulture/aquaculture-based professional skills depending on the subsystem to be serviced.

3.31 In the rainfed areas, differential extension strategies may have to be evolved for information-based technologies vis-a-vis skill-based technologies, which in turn could be either individual or group-oriented, whereas the present extension approach does not distinguish such subtle nuances. For instance, integrated pest management, moisture-augmenting rainfed farming, common pool resources management etc., call for mobilization of farmers into organized groups for self-governed action and taking control of their own destiny.

High Value Commercial Agriculture

3.32 Increasing population pressure directly impacts on future foodgrain demand. The faster economic growth phase that the country is likely to enter due to the current macro economic reforms will increase per capita incomes, which in turn, will not only raise demand for foodgrains but will also change the structure of demand for other products, i.e., non-staple crops. As a consequence, opportunities for diversified agricultural growth will expand.

3.33 Already a few pockets of commercialized agriculture have begun to emerge in India. Akin to the industrialized agriculture, the production systems are geared to cater to emerging markets—domestic as well as export. High value fruits like grapes, apples, etc., high value vegetables, floriculture, organized orchards, organized aquaculture units, sugarcane, oilseeds are some of the examples. However, if growing demands are to be fulfilled and if faster growth is to be attained in the medium and longer runs, there must be supply of newer technologies to improve the productivity, returns to crop, land and water management. Table 3.3 highlights the technological gap in selected high value commercial-oriented crops in India. Many of these crops suffer physiological and pathological disorders that defy technological solutions. Root wilt of coconut, rhizome rot of ginger, virus-like infestations in vegetables, 'katte' disease of cardamom, citrus decline, mango malformation, etc. are some of the examples of such disorders.

Similarly, post-harvest processing, quality control, storage, packaging and transport of high value crops have not received adequate attention.

3.34 High value crop growers fall into two groups. A majority of large producers have highly developed technical skills, information processing ability, managerial competence, market responsiveness and entrepreneurial skills, that foster efficient production systems.

TABLE 3.3 : **Technological Gaps in a Few Selected High Value Crops**

(in tonnes/ha)

Sl. No.	Crop	World Average	Highest	India Average	India Experi-mental
1.	Cabbage	23.10	70.9 (S. Korea)	6.05	40.43
2.	Cauliflower	13.26	26.7 (Australia)	7.38	27.30
3.	Garlic	5.55	31.36 (Egypt)	3.72	8.70
4.	Peas	6.35	14.82 (Belgium)	2.83	8.00
5.	Tomato	23.51	247.62 ((Netherlands)	9.78	40.45
6.	Onion	14.18	46.97 (U.K.)	10.60	40.00
7.	Potato	14.83	44.45 (Netherlands)	14.82	39.50
8.	Black Pepper	0.52	1.57 (Malaysia)	0.26	2.40
9.	Cardamom	–	0.3 (Guatemala)	0.06	0.35
10.	Ginger	–	–	2.18	8.00
11.	Coconut	–	5.06 (Malaysia)	1.00	4.70
12.	Tea	0.85	3.66 (Zimbabwe)	1.62	4.00
13.	Coffee	0.58	1.96 (Sri Lanka)	0.88	1.20
14.	Mango	–	–	8.80	NA
15.	Citrus	–	30.0 (Spain)	8.00	NA
16.	Pineapple	–	60.00 (Hawaii)	14.0	105.80
17.	Banana	–	–	15.5	174.00
18.	Grapes	16.75	23.43 (India)	23.43	60.00

Source: FAO Production Year Book, Vol. 39, 1985. IARI and CPCI Bose (1986).

This group has the capacity to demand requisite extension advisory services and the ability and willingness to pay for such services. On the other hand, around major metropolitan cities and towns, peasant-oriented, intensive but market-responsive high value vegetable cultivation and floriculture have emerged. However, unlike the organized group, the latter group is unorganized and relatively low skilled and will continue to depend on public extension systems. Exceptions to the rule are also abundant. For instance, exotic vegetable growers around Delhi have come together to hire professionals on a retainership basis. Similarly, small growers of grape, poultry, angora rabbits, mushroom, have also joined collectively to hire technical advisory and plant/animal protection services. In such cases public extension services may assume the role of a facilitator and catalyst to promote diversified agricultural systems. The public research system will still be shouldering major responsibilities of technology development and, on a selective basis, providing consultancy services.

4

AGRICULTURAL TECHNOLOGY DEVELOPMENT AND DISSEMINATION SYSTEM

4.01 India has one of the largest formal agricultural research and extension systems. They are complex, both in terms of institutional arrangements and organizational management. Their strategies of technology generation and dissemination have been top down and centrist (Biggs, 1989), i.e., technical and institutional innovations emanate as a result of systematic work of research institutes and get passed down in a linear fashion through extension to farmers. These centrist strategies have yielded rich dividends in the past, in the form of Green Revolution technologies and the self sufficiency in foodgrains production, which grew at about 2.7 per cent per annum and outpaced consistently the population growth. Yet there is an increasing sense of unease that the Indian agricultural research and extension system has lost its early vigour and is not in a position to respond on real time basis to the newer and emerging challenges. Of late, non-governmental organizations (NGOs) and private sector have also taken active interest in agricultural technology development and dissemination process.

NATIONAL AGRICULTURAL RESEARCH SYSTEM

4.02 The Indian National Agricultural Research System (NARS) comprises the Indian Council of Agricultural Research (ICAR) and the State Agricultural Universities (SAUs). Although agriculture is a State subject, the Federal Agency, ICAR is the central agency for coordinating, directing and promoting, agricultural research and education in the country. The council through the Department of Agricultural Research and Education (DARE) in the Ministry of

Agriculture links itself with the Central and State Governments and international bodies. The ICAR, besides having its own research and training institutions, supports through its central coordinating and financial grant making mechanism, the SAUs to discharge agricultural educational and research responsibilities of their respective jurisdiction.

4.03 The ICAR has a network of research institutions which provide basic and strategic research support and buttress the SAUs activities. This consists of 45 Central Research Institutes (CRIs), 9 Project Directorates (PDs), 20 National Research Centres (NRCs) and 71 All India Coordinated Research Projects (AICRPs).

4.04 The CRIs and NRCs have a national character and mandate. Increasingly, it has been recognised that the CRIs and NRCs should intensify their efforts on strengthening the basic and fundamental research (GVK Rao Committee Report, 1987), which has remained very weak in the Indian agricultural research. In the applied stream these institutions may only undertake strategic research programme, since they cannot be expected to develop technologies for heterogeneous farming situations or recommendation domains. As a corollary, gradually applied and adaptive research work will be increasingly undertaken through the decentralized structures available at SAUs and their Zonal Research Stations situated in different agro-climatic zones.

4.05 Major criticisms of applied agricultural research in India has been its repetitive nature, duplication of research efforts and overlaps especially in the applied and adaptive areas. G.V.K. Rao Committee identified six technical, administrative and operational problem areas in the Indian agricultural research which "threaten the very concept of evolving improved technologies". They are: (a) an inadequate focus on local problems in research programmes; (b) an excessive emphasis on uniformity of experiments and a straight-jacket approach in research; (c) inadequate financial resources and delays in the release of funds; (d) sub-optimal utilization of resources; (e) a lack of effective monitoring, and (f) weak links with extension. There is sufficient justification for rationalization, logical division of labour amongst the hierarchy of research institutions and deriving synergies by bringing about closer productive and meaningful interactions between CRIs/NRCs and SAUs. It is imperative not only to reduce duplication and unwarranted competition with similar work done in SAUs but also design a research management process that aids the researchers in striking a judicious balance between supply-pushed and demand-pulled research agenda and research prioritization; and ensures optimal use of scarce resources - financial and human.

Role of Zonal Research Stations

4.06 SAUs, before the advent of the NARP, concentrated their research efforts at the main campus, with inadequate attention to the technological needs of different agro-ecological regions of the State. NARP rectified the apparent weakness by strengthening and developing Zonal research capability of the SAUs and shifting focus from the main experiment station at the central campus to the regional research stations. It also enabled the SAUs, to create an agro-ecologically and geographically coherent and pragmatic basis for farmer-oriented research and develop effective research-extension linkage.

4.07 Concomitant with the NARP, the responsibility of regular training of middle level extension workers and subject matter specialists, was transferred to the SAUs in general and more particularly to ZRSs, which is best qualified professionally to conduct farmer-oriented adaptive research for generating technology and providing extension recommendations.

4.08 For improving the quality and relevance of research and training programme it was imperative that at ZRS, research programmes reflect not only the extension services agenda, but also research needs from the farmer perspective. The concept of status report was implemented, wherein, research scientists and extensionists of an agro-ecological zone could obtain a clear idea of the natural resources of their zone, the major crops and farming systems and the production constraints.

4.09 A typical status Report contains a description of natural resources and land-use patterns. Statistics on acreage, production and productivity of agricultural crops, horticulture, forestry and livestock, as well as human resources in the agro-ecological zone is sifted from existing soil, vegetation, land-use, climatic and irrigation maps and supplemented with statistical information from several State Bureaus. Based on the natural resource situation, combinations of major physical features such as soil type, water availability, altitude, temperature and crop growing period is used to delineate "farming situations" in which farmers grow crops and adopt cropping patterns with a very high degree of uniformity.

4.10 Farming situations are then used as major strata, wherein, joint surveys are conducted by the ZRS scientists and extensionists. These surveys reveal farmers resource utilization pattern, crop specific agronomic practices, and production constraints. It helped in a broad understanding of the farmers' environment and farmers' perceptions of the problems.

4.11 In addition to Status Report, several other mechanisms also enable researchers to access farmers' perceptions. These involve:

(a) Zonal Research and Extension Advisory Committees (ZREAC), which include zonal researchers, extension professionals and a few progressive and communicative farmers;
(b) monthly joint visits of researchers and extensionists and close interaction with farmers;
(c) diagnostic teams comprising researchers and extensionists visiting farmers' fields twice or thrice during the crop season to diagnose current field problems and provide solutions; and
(d) farm trials designed to validate technologies in farmers' field under their resource and management conditions.

4.12 The ZRS was expected to perform the following functions:

(a) strengthen decentralization of research;
(b) identify production or adoption constraints;
(c) pursue new adaptive research aimed at relaxing production constraints in the crops and crop sequences that dominate the zone; and
(d) verify research results from elsewhere (Bredero, 1988).

While there is little doubt that NARP certainly strengthened the research capability (for crop sub-system) of the SAUs, the team found that the research programme planning process at many ZRS, were neither relevant to the farmers needs nor the researchers have been able to embrace problem solving adaptive research mode. Several flaws observed in the earlier main experiment station based research still persist. Some of these were:

(i) Many researchers have yet to appreciate the concept of regionalisation of research. As a result, research method and projects are replication of efforts at main experiment station. For instance, it is difficult to justify setting up a tissue culture laboratory at the ZRS in Uttar Pradesh, or in West Bengal, where farmers have switched over to intensive vegetable cultivation in a particular zone, while at the ZRS no experiment was being conducted on production problems of vegetables, or a sole junior plant breeder continuing trials on the maintenance of germplasm of 18 crops in Uttar Pradesh;

(ii) The status reports by and large are descriptive and serve only as statistical compendium. Generally they lack an analytical approach. Nor do they provide insights into the different farming situations, farmers' perception of the problems, etc. Even in case of relatively good quality status reports, the team did not find its use in problem definition, research prioritization, and programme planning for farmer-relevant technologies;

(iii) Institutional mechanisms, like ZREAC, monthly joint visits of researchers and extensionists to farmers' field and joint diagnostic visits, that provide for effective farmer-researcher interactions, whenever took place were not effective enough to adequately influence the research agenda. In some instances, such as in Assam even though researchers had developed better perception of the farmers' problems, neither existing projects could be altered nor newer research initiated to provide succour to farmers' problems. It appeared that ZRS has no access to flexible funds, as NARP made provisions only for civil works to create infrastructure, equip the ZRS and provide incremental staff with a contingency fund of Rs. 20,000 per incremental scientist per year. This amount is inadequate to meet even the operating cost of ongoing on-station research. Most of the staff and facilities are tied up with ongoing projects. Weak research planning and management process exacerbates the problems of weaker institutional base.

(iv) Most of the ZRS is expected to service heterogeneous and large number of farming situations/micro ecological niches, that is spread over 4-6 districts. However, many of these niches could not be simulated at the experiment stations of the zone. Neither the ZRS has facilities to conduct on farm research nor the researchers are equipped to undertake farmer participatory approaches of problem solving. As a result many problems remained unattended;

(v) In yet another ZRS in Uttar Pradesh Junior Researchers were assigned the responsibilities of servicing the T&V personnel during monthly workshop. In the absence of significant research and field experience, they were not able to satisfy SMSs' problems. Further, the linkage with Krishi Vigyan Kendra is very tenuous. Even in certain regions of Uttar Pradesh where ZRS and KVK are housed at the same site, the institutional mechanisms for closer collaboration was begging question;

(vi) Only a few ZRSs are fully staffed with strong leadership. Even in the adequately staffed ZRS, scientists composition bear no correspondence with either the existing or emerging sets of production problems. It is worth repeating the ICAR review finding: "There is a dearth of scientists in the fields of water management, animal husbandry, agricultural economics, agroforestry, agricultural implements, and post-harvest technology (ICAR, 1989)"; and

(vii) The inappropriate skills-mix at ZRS, an imperfect understanding of the basic purpose of zonal concept of agricultural research as well as non-acquaintance with farming systems research perspective and farmer participatory research, lack of flexible funding and mobility are among the most important obstacles to conduct of farmer participatory research.

AGRICULTURAL EXTENSION

4.13 Agricultural extension, unlike research, is entirely a State Government responsibility. It is almost eighteen years since India initiated substituting its extension system, based on community development era, that employed multipurpose village development workers, by the Training and Visit (T&V) method of agricultural extension.

4.14 T&V method was expected to overcome the deficiencies observed in the earlier extension service, such as:

(i) long and multiple lines of command;

(ii) field level extension workers devoted less than 25 per cent of their time in agricultural work instead of the prescribed 80 per cent because of the multiple demands of other duties–public health, family planning, savings groups, statistics, etc.,

(iii) almost no technical advice was being given in the field and most of the extension workers' time was consumed in input distribution and reporting;

(iv) Research was academic in nature and unrelated to the farmers' real problems;

(v) Extension staff were in great need of further training, particularly in laying out field demonstrations.

4.15 The subsequent reorganization (Benor and Harrison, 1977; Benor and Baxter, 1984) of extension in the form of T&V system involved the establishment of a unified hierarchial service and a single line of command from the State Directorate of Agriculture (DOA) to the field level extension worker. The latter was relieved of all duties except agricultural work and exclusive focus was brought on extension. A time bound method of managing extension was introduced that included a manageable staff-farmer ratio, a fixed schedule of properly supervised visits designed to deliver selective and timely impact messages on technology with strict regularity, explicit field orientation – the village staff is not even provided an office, minimal paper work and administration. In principle all extraneous activities such as those connected with the physical handling of inputs, credit applications and implementation of other developmental programmes were to be strictly avoided. The system sought to encourage productive links with agricultural research institutions through regular intensive training sessions for Subject Matter Specialists at monthly workshops, followed by fortnightly training sessions for village extension workers (VEWs). Extension agents were expected to liaise with other departments, input supply agencies and to advise farmers on sources of supply. The VEWs during the village visits, were supposed to focus mainly, but not exclusively on pre-selected contact farmers – 8 to 10 in a village. Over a period of time, an increasing number of follow-up farmers were expected to learn from contact farmers the advice, translate them into adoption behaviour and gradually attend meetings between

VEW and contact farmers (Feder and Slade, 1989).

4.16 In a major review of the implementation of the T&V method of extension in India, the World Bank's Agricultural Extension Sector Review (1987) acknowledged that the method was institutionalized in most of the States and had a positive impact on agricultural performance. However, it also identified a number of problems both internal to the extension service and challenges originating outside the system which needed institutional responses to ameliorate the weaknesses in the reform process. The internal issues raised were the role, deployment and quality of field staff, including SMSs; the links between extension and research; the place of women in extension activities; the responsibilities of DOE in the MOA; and improved M&E. Subsequently, Government of India appointed a screening committee under the Chairmanship of Mr. K. N. Ardhanareeswaran, the then Additional Secretary (Agriculture) to vet the findings.

4.17 Ardhanareeswaran Committee (1988) accepted all the findings of Sector Review except one that alluded to the responsibilities of DOE. It maintained that the DOE should retain its role of coordination, extension development and providing implementation assistance to States. However, the committee conceded that the extension, hitherto, narrowly focussed on cereal crop production must be broad based and embrace a farming systems approach; that the extension staff need to appreciate farmers' experience while modifying the impact of messages to suit location specificities; that the implementation of special crop development schemes should not be allowed, as far as possible, to disrupt the rhythm of training and visit schedules of VEWs; that the SMSs remained the weakest link in the chain, and that the administrative chores took precedence over technical roles at higher level SMSs (above sub-dvision level); that the technological recommendations were yet to be adapted to the agro-ecological requirements; that in the absence of suitable farming technologies, the extension services were hampered in the rainfed region; that the extension services continued to be gender blind; and that the in-house M&E units have not displayed their capacity to generate management-friendly evaluation reports for mid-term corrections.

4.18 During the visits to different States, the team encountered several versions of implementing T&V Obviously, at the State level the policy makers and agricultural administrators have been responding to different sets of compulsions. The different implemented versions are:

(i) *Exclusive focus on extension* : Directorate of Agriculture has created parallel lines of command to implement T&V system of extension and other special development programmes like National Watershed Development Programme, National Oilseeds Development Programme etc. However,T&V system in this case does not influence the design or implementation of other development programmes. This type of implementation can be found in Andhra Pradesh and Himachal Pradesh;

(ii) *Multiple task but single line of command* : In this case Directorate of Agriculture at the State level has several Additional Directors of Agriculture overseeing different areas like T&V, inputs, development programmes, etc. However, all these tasks are integrated and implemented through a single line of command at zonal level through VEW. The VEWs have specific targets to achieve on different areas of agricultural development programmes that are either sponsored by Central or State governments. In this case, which can be observed in most States, VEWs claim that administering the programmes and filing returns take precedence over extension responsi-bilities;

(iii) *Parallel lines of command but multiple responsibilities for the VEW of T&V*: In Tamilnadu and Uttar Pradesh, it was observed that the Directorates of Agriculture have several activities on board besides T&V and implementation of special development programmes, such as seed production and distribution, distribution of plant protection chemicals, etc. While the input division consisting of seeds and plant protection chemicals has parallel structures right from the State to the depots that operates as retail outlets at the grassroot level, the extension VEW is assigned the task of motivating farmers to offtake seed and chemicals, and specific target is also set. The special development programmes, both Central and State sponsored, are also implemented through the VEWs. In Uttar Pradesh the subsidies on seeds are available only through the departmental depots. Consequently, Tarai Development

Corporation's (TDC) marketing channel is not only defunct, but it has transferred the seed distribution responsibility to the DOA network. The TDC as a result has lost its touch with the clientele and produces seeds that may not be required by the farmers. A common refrain amongst the farming community of the Tarai region and the irrigated plains of Western Uttar Pradesh about poor seed quality, is not effectively feedbacked to the seed agency. Credibility of the hapless VEW, who pushes the seed, is eroded consequently. Private agencies feel discouraged to enter the State seed business as seed subsidy is available only at State retail depots rather than at source. In well endowed areas of Uttar Pradesh and Tamilnadu, continued role of DOA in input distribution can be discontinued and both seed corporation and plant protection chemicals manufacturers allowed to open their respective outlets.

4.19 The proponent of the latter two options argue that unless extension messages are backed up with special crop schemes, agricultural implements or other development programmes production and productivity cannot be increased. A key feature of these schemes (National Oilseeds Production Programme, National Pulse Production Programmes, Special Food Production Programmes) is the concerted effort to promote a packaged programme in the mission mode and provision of subsidies on key inputs like seeds, fertilizer and credit for crop demonstration. Rajasthan, which is a representative of second typology has mixed the T&V system and special schemes without burdening the VEW substantially (Box-1).

BOX - 1

Judicious Balance Between T&V and Special Schemes in Rajasthan

Rajasthan DOA has been innovative in utilizing both GOI and State Government funded programmes to enhance the quality of transfer of technology. It has about 4000 VEWs, who discharge both the TOT and programme implementation responsibilities in 30,000 villages. In a simplified and single command structure, the DOA has a presence below the sub-district level in Rajasthan in the form of AEOs and VEWs of the T&V system. It has actively sought to privatize distribution of inputs – seed, fertilizer and pesticides — by delicensing dealership at the grassroots. However, DOA is directly involved in the distribution of agricultural implements, plant protection equipment, gypsum and bio-fertilizers that are subsidised either through State or Centrally-sponsored programmes. VEW only certifies entitlement of other subsidised inputs which farmer can claim from the cooperative retail outlet on purchase.

In 1990-91, for instance, the programme related expenditure (Plan Outlay) was doubled to Rs. 16 crores without increasing staff size. The enhanced Plan Outlay envisioned:

(a) A four-fold increase in the distribution of Plant Protection equipment from 8000 to 32,000 : (direct handling by VEWs);
(b) Distribution of bullock drawn agricultural implements from 3500 to 9000: (direct handling by VEWs);
(c) Distribution of sprinkler sets increased from 2000 to 3500: (VEW has a physical verification role);
(d) a 10 lac ha. increase in area covered under Plant Protection measures from 60 lac ha to 70 lac ha (VEW has a motivating role);
(e) a 26 per cent increase in certified seed sales from 1 lac qtls. to 1.30 lac qtls: (VEW has a motivating role);
(f) a 26 per cent growth in fertiliser sales from 2.90 lac tonnes to 3.60 lac tonnes: (VEW has a motivating role).

These distributions were possible because of the following special schemes:

Schemes	*Outlay*
Oilseeds Production Programme	Rs. 600 lacs
National Pulses Development Programme	Rs. 250 lacs
Special Food Production Programme	Rs. 400 lacs
State Programmes	Rs. 500 lacs
Total	Rs. 1,750 lacs

On a crude calculation, the outlay per VEW is Rs. 43,750 in a year. However, it is important to note that a lion's share is allocated as subsidy at source – State Seeds Corporation for subsidising the cost of all certified seed in the State, sprinkler suppliers and Cooperative retail outlets for distribution of subsidized pesticides, where the VEW has a certifying role only. Therefore, a VEW may be handling less than Rs. 10,000 per annum. The workload of a VEW was increased on an average by:

(i) direct distribution of plant protection equipment–8 Nos. Bullock drawn implements – 2 Nos.
(ii) physical verification of the installation of one sprinkler for ensuring subsidy entitlement,
(iii) motivating a few farmers to adopt efficient water resource management and simultaneously adopt inputs like improved seed, fertilizer and pesticides.

4.20 Since most of the States are resource starved, they accept the centrally designed and sponsored schemes which are always flushed with funds. The State DOAs in general do not attempt to repackage the schemes to be congruent with the regional and local needs. However, Andhra Pradesh DOA took initiative in utilizing the special schemes to the advantage of technology adaptation and resource allocation to areas with larger potential marginal contributions (See Box 2 and 3).

BOX - 2

Linking Special Crop Schemes to Technology Transfer

Centrally Sponsored Special Food Production Programme (SFPP) followed the erstwhile Intensive Agricultural Development Programme approach of the 1960s. It aimed at reversing the declining trend of production of cereals by targeting of beneficiaries in selected favoured areas by means of subsidies on key inputs, credits and crop demonstrations.

DOA in Andhra Pradesh in 1987-88 carried out a regional analysis and discovered that the most favoured areas, i.e.,

Krishna-Godavari, North Coastal Zone and Southern Zone, endowed with high proportion of irrigation (42 to 60 per cent) and adequate rainfall 700-1100 mm. in general had adopted the miracle seed based technologies of rice. While scarce rainfall zone of Rayalaseema, high altitude and Tribal area, with 15.7 and 13.3 per cent of irrigated area hardly could achieve higher productivity and production in rice. A rice productivity analysis helped in differentiating mandals into high more than (2.5t/ha.), medium and low (less than 1.00t/ha.) productivity mandals. Mandals experiencing high and low productivity were eliminated, while Mandals falling in medium productivity range were subjected to further micro analysis to determine the production constraints and availability of water resources. Potential mandals were selected and the subsidies available through the SFPP were used to bridge the gap between recommended technological package and prevalent farmers' practice. That is, instead of covering total inputs with subsidy, it was converted into incentive, which could be availed by farmer only if he adopted non-monetary agronomic recommendations that were identified as critical.

The SFPP could thus be used to encourage the disadvantaged farmers to increase production while simultaneously linking the programme to effective adoption of improved non-monetary package of technology.

BOX - 3

National Watershed Development Programme Fosters Onfarm Research in CDR Agriculture

National Watershed Development Programme for Rainfed Agriculture (NWDPRA) is a critical response of federal government to the developmental needs of dryland agriculture. Paradoxically, it is also a resource rich programme for the resource poor regions. However, like all top down development schemes NWDP also had a packaged scheme that did not allow for manoeuvring to address local needs. Nor it has a line item in its budget for encouraging adaptive research to validate and fine tune dryland agricultural technologies to farmers' context.

In Andhra Pradesh, Dr. N. K. Sanghi (a dryland agricultural scientist), who had been working closely with farmers and NGOs, even before "farmer-first" approaches became fashionable, was confronted with the dilemma of responding to farmers' production problems without access to flexible funds in 1987. As an agricultural researcher he was trying not only to blend indigenous knowledge with formal experimentation, but also forge a consortium of NGOs to tackle problems of soil and water conservation, design farmer relevant integrated pest management strategies of Red headed caterpillar and root grub of castor and groundnut respectively. These problems were not receiving adequate on-station research attention. But Dr. Sanghi and his collaborators had no access to research funds as the research station funds were tied up with ongoing academic experiments.

The then Agricultural Commissioner and Secretary, Government of Andhra Pradesh came to their rescue. He had also been receiving several dryland related problems from his field level functionaries, that the researchers were not able to respond. The Secretary provided the NWDP funds for adaptive research provided the extension services' field based problems were also solved. Dr. Sanghi and his NGO collaborators got a contractual funding to undertake collaborative research with extensionists and farmers to solve farmers' most immediate production problems. Based on this successful experience, when the NWDPRA guidelines were revised in 1990, 'onfarm' research became a part of the Project design.

4.21 A general disillusionment with the concept of "contact" farmer was observed across the T&V hierarchy. Contact farmers are usually selected by the extension personnel and are supposed to be representative of socio-economic groups that prevail in a particular village. However, in practice, the study team found that invariably local elite pressures and staff preferences bias selection in favour of the richer and more influential farmer. The extension hierarchy in different States felt that such contact farmers neither transform extension messages into actual practice nor they are interested in transmitting technology to follow up farmers. In such a situation farmer participation cannot be fostered.

4.22 The public agricultural extension system in several States have started experimenting with alternate approaches to increase the efficiency and effectiveness of extension through farmers groups (Box 4).

BOX - 4

Public Agricultural Extension Goes Group Based

(i) In Assam, DOA encourages farmers of a village to group into *Pathar Parichalana Samity* (Field Management Committee). Agricultural Extension Officer (AEO) alongwith the Village Level Extension Worker (VLEW) mobilizes farmers of contiguous field into *Pathar Parichalana Samity* (PPS). Actual tillers of the land-owner farmer and tenant farmer of a contiguous field (block) join together and apply for registration of a local PPS with the District Agricultural Officer. The AEO or VLEW acts as a technical adviser to the PPS and enables the group to acquire newer technical and organizational skills.

In Lakhimi PPS of Naogaon District, the AEO and VLEW initially replaced the contract farmers with the executive members including President, Secretary and Treasurer as the contract group. The contract group was expected to transmit technologies to the rest of the PPS. However, this group also proved ineffective. The PPS General Body suggested that it would nominate the contact group farmers' field, however, would be used for minikit trials, demonstration and organizing field days.

In addition to membership fee, the village youth contributed wages earned in implementing a social forestry activity on the village commons. This was used for evolving social fencing and self-governance. Subsequently, one more PPS was formed in the village.

The AEO responded to non-crop technological demands of the PPS by liaising with other line departments, such as Animal Husbandry, Fisheries, Irrigation and inviting the concerned officials to address the PPS. The farmer group may form the basis for broad basing extension.

The PPS has also evolved an innovative input acquisition system to back up technological messages into adoption. A few members of the Executive Committee form the input procurement group. This group assesses the members' demand for seeds, fertilizers etc. and procures the same from input dealers in the district, thereby ensuring timely availability at the farm gate and deriving economies of scale through centralized operation. The input price is inclusive of the transportation cost

(ii) In Rajasthan, DOA inserted press advertisement in the vernacular press inviting farmers, desirous of technical assistance, to form voluntary groups of atleast 20 farmers. The DOA provides training to the Executive Committee on leadership and organizational aspects. The individual contract farmers are being replaced by such voluntary farmer groups. This experimentation is still in its infancy.

(iii) In Kerala, the public extension system has broad based its focus through, now famous, group farming. The technology dissemination and adoption takes place through the specially created farmer groups.

4.23 The methodological rigidity and its doctrinal implementation also drew adverse reaction from principal consituencies of the public extension system. Initial enthusiasm about fixed and supervised visits, continuous trianing has given way to monotony and ritualistic implementation. In a centralist mode (Biggs, 1989; Biggs and Farrington, 1991) technology messages pass from researcher-cum-master trainers to Subject Matter Specialists to VEWs to farmer. However, technology innovation process is complex, but is not changing as fast as the monthly and fortnightly training. Frequent repetition of the same message all through the passage without being sensitive to the clients' needs has robbed the vitality of the system. The entire process has become too ritualistic. Feedback from farmer to researcher through extensionist remains the weakest link.

4.24 M&E units have been well established. M&E has, so far, focussed its attention on the quantity rather than quality. Thus, the number of VEW—farmer contract, its regularity, the extent to which the impact points are adopted or not adopted and the yield measurements are the parameters of monitoring. It neither monitors farmers' indigenous knowledge, nor knowledge of recommended practices. It does not go beyond the binary classification of adoption and examine the rationale for non-adoption and the ongoing adaptation/experimentation that the non-adopted farmers may be carrying out.

The M&E units are yet to embark on evaluation of T&V experiences and create a learning system that is capable of self-correction.

4.25 The M&E findings are seldom used towards either providing incentives/rewards to the high achievers or reprimand low performers.

4.26 In the absence of institutional mechanism for performance rewards and incentives and lack of career advancement, the AEOs and VEW have lost their enthusiasm and are demotivated. There is no premium attached to individual initiatives.

4.27 To broad base agricultural development in the country will mean besides enlarging coverage of non-crop enterprises also developing a range of educational programmes that can transfer appropriate technology and develop technical, managerial and human development skills to all types of farmers on a more equal basis. However, subsistence level women farmers, inclusive of female headed farm households, and young farm families, who have been outside the ambit of extension, have special technological and educational needs. Broad basing should also imply that these two disadvantaged clientele and their needs are specially covered. That is, the extension service should become gender-sensitive as well as youth sensitive.

RESEARCH–EXTENSION LINKAGES

The existing structure of research-extension linkages comprises:

4.28 A national level Research and Extension Coordination Committee headed by Director General, ICAR with representation from ICAR, Ministry of Agriculture, Planning Commission, NGO's etc. for the purpose of coordinating research and extension functions at the central level and developing policy guidelines in this regard.

4.29 A State Level Research-Extension Coordination Committee chaired by Agricultural Production Commissioner and comprising the Vice-Chancellor of SAU's Secretaries and Directors of allied departments, representatives of NGO's, farmers and others to review functioning of research-extension coordination mechanisms at the State level and consider issues arising out of ZREAC for formulating policy guidelines.

4.30 A Zonal Research and Extension Advisory Committee for the purpose of operationalising research-extension linkages. This Committee is established at the ZRS level in each agro-climatic zone and is chaired by Director Research/Extension of the SAU. It consists of the Additional Director (Extn.) and of Zonal Joint Directors, Department of Agriculture, Zonal Coordinator, Directors, State Departments or their nominees, Heads of KVK's/FTC's, heads of sub-stations. This Committee is to meet pre-seasonally to identify technological and socio-economic constraints in transferring location specific technologies and develop an action plan for the same, to review feedback from extension and the solutions offered to the same by the researcher, to plan the schedule of monthly workshops and visits of the joint diagnostic teams for the forthcoming season and to practice and review the conduct of on-farm trials for the zone.

4.31 Monthly workshops, wherein, the Master Trainers from the ZRS train the SMS of the Agricultural Departments. This mechanism serves as the clearing house, resolve problems and issues faced by extension and research system and also planning the extension message for the coming month.

4.32 Joint diagnostic visits by teams of scientists from the ZRS and Extension Officers from the State Department once during each season.

4.33 Almost all the formal linkage mechanisms have by now become ritualistic and least effective in ensuring that farmers' problems are reflected in the ongoing research/extension activities. The concept of diagnostic visits and feedback is not clearly understood.

NON-GOVERNMENTAL ORGANIZATIONS.

4.34 Non-Governmental Organizations (NGOs) have emerged as a significant "third sector" with dedication, capacity and commitment to overcome the shortcomings of the public and the private sector in poverty alleviation and development activities. NGOs have advocated "bottom-up" approach involving rural poor directly in design and implementation of both income-generating and social activities, and have "empowered" them to develop on their own.

4.35 In India, NGOs can be distinguished according to the nature of their relationships to the rural poor, and their staff composition. Membership organizations (Caroll, 1993) are staffed and elected by the people they are supposed to serve and represent, e.g. farmer organizations. Non-membership Organizations are, on the other hand, staffed by professionals. In addition to undertaking particularly agricultural technology development and dissemination the latter usually build and support membership organizations in the empowerment process. NGOs like MYRADA, AKRSP, BAIF, Deccan Development Society, Action for World Solidarity (ASW) that belong to the second type category, have demonstrated potential for closer collaboration between NGOs and the public agricultural research and extension services in the operationalization of participatory technology development and dissemination as well as strengthening the latter's effectiveness in improving livelihoods for the rural poor. Whilst, public research and extension systems view technology development graduating in 'stages' viz. from basic through strategic, applied and adaptive research, with emphasis on agricultural technology development – dissemination linkages; NGO's agricultural technology development is "issue" based that more often integrates several stages simultaneously in an action research mode.

4.36 The NGOs have been particularly innovative and relatively successful in the following major areas: (Bebbington and Farrington, 1992):

Participatory Methods of Diagnostic and Farming Systems Approach to Agricultural Technology Development

- MYRADA, Seva Bharathi, Ramakrishna Mission have successfully developed and adapted participatory rural appraisal methods, agroecosystem analysis procedures and designed training programme for both other NGOs and Government staff. (Fernandez, 1991; Mandal *et al.*, 1991).
- Action for World Solidarity established a network of

grassroot organizations in Andhra Pradesh and collaborated with public research/extension system in designing an integrated pest management strategy for controlling red hairy caterpillar (*Amscata* sp) of castor. Subsequently, ASW was instrumental in formation of farmer organization for collective action in the selected villages to achieve maximum impact of group-based technology. (Satish *et al.*, 1991; Sanghi *et al.*, 1992).

- Ramakrishna Mission, Narendrapur adapted a whole farm approach in blending informal and formal agricultural research in the coastal ecosystems of West Bengal. It helped existing rural youth clubs to mobilise local resources in integrating crop-aquaculture technologies based on the indigenous practices of land levelling in swarm marshy areas. (Das *et al.*, 1992).

Innovations in Technologies and Management Practices

- The Bharatiya Agro-Industries Foundation (BAIF) initiated research into frozen semen technology and has been responsible for producing around 10% of the country's cross-bred dairy land through its 500 field units in six States. It has expanded its programme to address animal health and nutrition aspects. It also participates in several of the Indian Council for Agricultural Research – sponsored All-India Coordinated Research Programmes (Satish and Farrington, 1990).
- Society for Rural Industries has developed land and water conservation technologies in the tribal plateau of Chotanagpur region.
- In the adaptive end of the R&D spectrum, PRADAN has successfully scaled down technologies developed at public institutes for mushroom and raw silk production. In the case of leather processing, besides adaptation of technology it has also devised integrated schemes of credit and marketing (Vasinalai, 1991).
- Society for Hill Resources Management School has successfully evolved multi-tier plantation technologies and has been instrumental in introducing a social organisation and management dimension for arresting environmental

degeneration process in Bihar (Mishra, John, Bottrall, 1992).

Dissemination Methods

Several NGOs have sought to develop horizontal dissemination methods. For instance, farmer-to-farmer spread of fish-based farming systems technologies by Ramakrishna Mission through the support of CAPART; soil and moisture conservation technologies by AKRSP through the mechanism of farm extension volunteers; sericulture technologies by Sri Aurobindo KVK through training and back-up support to rural master trainers are illustrations of innovative dissemination methods.

Improved Training Methods

The AKRSP identified village training needs through discussions with farmer organizations. Initially, AKRSP mobilised existing public provision (i.e. KVKs and FTCs) for training farmers. The classroom based courses were very formal and lecture oriented. Farmers' evaluations revealed limitations of conventional training methods. AKRSP, in response, developed an alternative training and dissemination methodo-logy of using innovative farmers as resource persons, which it tested at several locations. Official trainers, were, subsequently, brought as observer participants to facilitate methodology adoption. Successful adoption was reinforced by informal networks that fostered experience sharing and consultations (Shah and Mane, 1991).

4.37 The lessons emerging from the farmer participatory projects initiated by the NGOs are:

1. *Participation as a process*: Successful farmer participatory projects emphasised the process of participation as opposed to immediate quantitative outcomes. They also demonstrated flexibility and willingness to experiment.
2. *Balance between human resources development and transfer of technology*: Instead of adopting a narrow view of extension, farmer participatory projects ensured a balance between human resources development and transfer of

technology since both aspects of extension practice are important and indeed mutually supporting. Neither awareness building and education with no tangible objectives nor short-term tangible gain through transfer of technology with no substantial base for longer-term participation are satisfactory recipes.

3. *Institution building is a pre-requisite for farmer participation*: Farmer participatory projects were built upon a local organisation to ensure a secure and locally available foundation for future activities as well as minimising the external dependency. Institution building activities included mobilisation, organisation, training, technical support and system management. The concept of participation goes beyond joint experiment or on-farm trials. NGOs seek to strengthen farmer organizations so as to enhance the rural poor's capacities for self management and negotiation with external institutions and dominant interest groups. They have emphasized institution building methodologies that strengthen the coordination among individual growers and subsequently between communities.
4. *Transfer of Technology through Groups*: NGOs have been addressing problems in CDR agriculture that manifest common pool resources characteristics. Institution building process aids in group based technology transfer.
5. *A team approach*: A common feature of many participatory projects is the presence of a project team which ensures genuine client participation. External catalytic support is fundamental to the process; but reliability and regularity are critical. Given the complexity of the process of participation and the fact that in its initial stages at least it is labour intensive, a team approach has advantages over the lone field worker. A team allows both for the sharing of experiences and for the mobilisation of resources.
6. *Stake building*: Projects that promote participation have invariably designed experimentation with stake building exercises in the local organisation. It may take the form of membership fee, proportionate contribution from the wage income and usufruct rights, savings and thrift etc. Besides building a corpus for the local organisation, it encourages regular contact and discussions among local community as

to better ways of using funds and thereby provide continuity in situation analysis, local responses and stakeholders contribution. This process also enables owning up of the local organisation and promotes emergence of functional leadership.

5

RECOMMENDATIONS

I. GENERAL RECOMMENDATIONS

Reconceptualising Public Extension and Research as Integral Components of Agricultural Knowledge Information System

5.01 Instead of viewing public research and extension in isolation as the only source of technological and institutional innovation, it is useful to acknowledge that other sources of innovations also exist. Both NGOs and private sector have begun to play an important role in agricultural technology development. Moreover, farmers have been traditionally experimenting and innovating. Farmer-to-farmer diffusion has remained the major source of dissemination. Besides multiple sources of innovation, there is an array of providers rather than a single extension delivery system. The providers in the agricultural knowledge information system should be linked by two way communication channels between them (Figure 5.1).

5.02 Diffused responsibility for extension and research creates a pluralistic institutional arena which acts as a force for change in policy decision making. The impact of multi institutional extension and research activities requires that the public sector play an important role in coordinating and ensuring accountability of the agricultural extension as a whole.

5.03 The T&V system, for strategic reasons, initially focussed narrowly on technology transfer of major crops. But coping with the emergent problems of various agricultural situations calls for enhancement of capabilities of farmers, broad basing extension and providing extension coverage for disadvantaged segment of farming community, i.e., farm women and rural youth, and landless labour. Therefore, the goals of public agricultural extension system should be

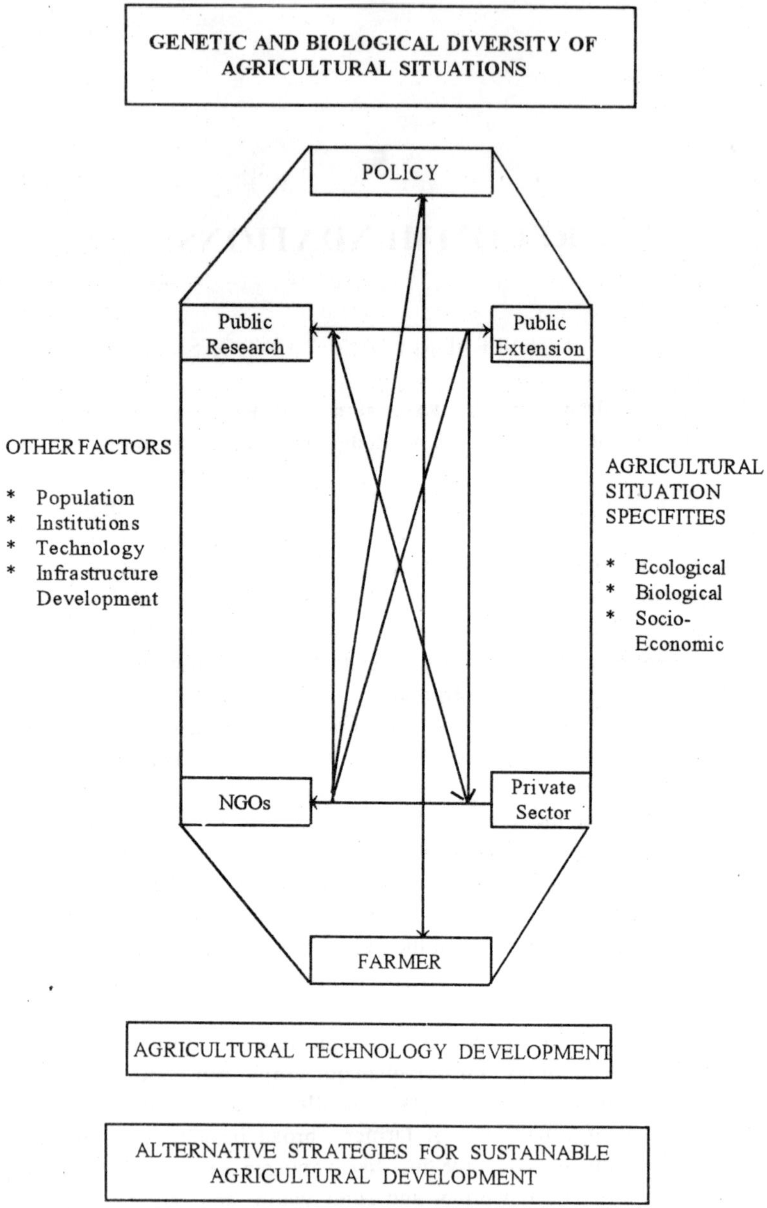

FIG. 5.1 : **Multiple Sources of Innovation**

enlarged to include, in addition to technology transfer, human resource development, broad basing and farming systems approach and gender-differentiated strategies within its agricultural extension policy.

Farmer Involvement in Agricultural Extension and Research Policy

5.04 Agricultural Extension and Research Policy should be consistent with and supportive of the recently announced national agricultural policy. Farmer involvement, through formal mechanisms and informal consultations, in policy formulation and periodic review is an effective means of creating "demand driven" national agricultural extension/ research system. It should be responsive to all major groups of farm people and flexible to encourage concerned public, private and nongovernmental organisations to contribute to the agricultural development goals.

Proactive Role of Clientele

5.05 No one extension approach is appropriate for all conditions, objectives or clientele. However, the basic principles of successful extension system may be common. Extension Programme priorities should be based on both an assessment of client needs and client participation in programme formulation. The client groups, as well as key collaborating agencies, should be involved in reviewing and providing feedback on extension programme content and approach as a means of improving performance.

Importance of Farmer Groups in Client-Driven Extension/Research System

5.06 User groups, including farmers' organisations, farmer clubs, commodity growers' associations, young farmer clubs, special interest groups and the like are effective institutional devices in the creation of a client-driven agricultural research and extension system. Extension can play an important role as a catalyst in organising these farmers groups. These groups then can become important mechanisms in articulating specific research and extension needs, in facilitating horizontal diffusion, in accelerating technology dissemination and in developing technical and managerial skills and knowledge of farmers who participate in these organisations.

5.07 It is also found that group rather than individual approaches provide a sustainable participation. Moreover, participatory approaches become empowering if extensionists tend to gradually handover most, if not all, of their managerial and organizational functions to self-governing groups. But world-wide experiences suggest that neither top-down structures nor external imposition can enable empowerment until and unless rural people enhance their capacities to articulate their needs and actually participate in the development process. However, since resource poor farmers are constrained in organizing themselves, external credible agencies can play the role of facilitator–assisting capabilities building and forming groups as a part of empowerment strategy and in response to the emergent needs.

5.08 It is well known that the poor and disadvantaged population would not be able to bear the transaction costs of organising, while existing groups may not be involving the poorer sections at all. They may also not have the necessary management skills. Some of other well known problems in organizing and working with groups in the class\caste driven Indian societies, such as vulnerability of domination by vested interests of elites, or the hazard of "free rider" problem, and internal democracy of such groups, are vexed and complicated. Moreover, the time, effort and skills required to form sustainable local groups capable of "owning" a process of change may not be available with public systems.

5.09 NGOs, on the other hand, usually work with groups and have the necessary skills of eliciting empowering participation. Some of the difficulties of organizing and strengthening groups may be eased if extension professionals can link with NGOs. The latter can facilitate on-the-job learning and skill transfer in the areas of group mobilization, institution building and human resources development to the farmer. However, the quality and distribution of NGOs is uneven, and NGOs will collaborate with public systems only where the perceived benefit outstrips the transaction costs.

Guiding Principles of Group Formation

5.10 Farmer groups formed around joint management of a natural resource, i.e., ground-water and associated problem of either seasonal or permanent draw down regeneration of degraded land and other common pool resources; around technologies that require joint management — Integrated Pest Management; around an economic enterprise — single commodity groups, especially concerned with input supply, processing, marketing; or around the testing of technological options have found to suffer, to a lesser extent with group formation problems cited earlier. However, extension and research workers will require flexibility and new skills, which in turn, require changes in their training, if they are to work successfully with a wide array of special groups. Creating large numbers of informal/formal groups, either on a "special interest" or more general basis is likely to be one of the biggest challenges for extension. The necessary switch in mind-set from "providing technology packages" to building opportunities for horizontal learning within the group or across the group through, for instance, innovators' workshops will be another.

5.11 In our review of existing T&V system, breaking down of contact farmers was reviewed in Chapter 4. Another issue that has surfaced is the cost of T&V. The T&V system of extension relies primarily on visits by VEOs to contact farmers. It envisages a major build-up of personnel in the extension services, both sub-professionals at local level for implementation, and professionals at regional level for supervision and monitoring. The concentration of effort associated with successful T&V therefore tends to be expensive. Efficiency and effectiveness of extension can be enhanced by working with farmer organizations and promoting Farm Extension Volunteers (FEV).

Farm Extension Volunteers

5.12 The local farmers organization can identify among its members a few potential innovators and entrepreneurs for acting as Farm Extension Volunteers (FEV). They can provide services which complement those made available by VEWs, and in places where VEWs are not positioned, they can discharge the VEWs responsibilities, thus greatly reducing burdens on the public budget and on its administrative resources and thereby increasing group self-reliance.

5.13 Village extension volunteers will need practical, job related training that should include technical, communications, and social skills. They may also require periodic reinforcement.

5.14 Such trained community members besides experimenting and practising improved farming systems also assume the role of disseminator. This participatory strategy of identifying and training large number of extension volunteers whose background is appropriate to local-needs can overcome the problem of large vacant positions. The FEVs may be either unifunctional or polyvalent. In any case, they should begin with a limited range of functions which can be expanded after the initial tasks have been mastered.

5.15 Extension volunteers also require supervision, support, and dependable sources of information and material inputs. VEWs role and skills should be gradually modified to take account of the increasing presence of extension volunteers. VEWs, thus, would require retraining to provide back-up technical and managerial support for the extension volunteers.

5.16 Farm extension volunteers are accountable to the local organization. While voluntary service may be an important motivation, village extension volunteer would usually require some modest material compensation, even when they serve on a part-time basis. Resources required in the form of a token honorarium to put the concept of extension volunteers into practice can be provided to local farmers organizations on a tapering basis. Additional help in mobilising loans for setting up service facility for volunteers, i.e., input distribution or custom hiring will enable them to turn entrepreneurs. Gradually, the farmers organization is expected to bear the cost of Volunteers' services. Thus, the Farm Extension Volunteers (FEV) would remain accountable to the local organization.

5.17 Instead of acting as a "delivery system", the public agricultural extension, henceforth, assists local people in organizing themselves to develop their own "acquisition systems". Thereby, the local community can acquire from the external environment those technologies and services that they need.

Participatory Agricultural Development

5.18 Farmer participation in agricultural technology development and dissemination is designed as sequences of activities during which technology is developed and disseminated, i.e., Diagnosis, Prioritization, Planning and Designing technological solutions, Implementation, Evaluation and Feedback, and Dissemination.

Diagnosis

5.19 Mapping of agricultural situations and inventorying of production problems is the first step. Participatory Rural Appraisal (PRA) techniques will help extensionist become acquainted with range, intensities, context, problems and opportunities. This step helps in knowing what we don't know but they (farmers) know. Farmers/Farmer organizations, Extension and NGOs are involved in the process.

5.20 Identification of homogeneous agricultural situations. Overlaying of maps, identification of patterns, analysis of partial use of land over time and the resource flows and farmers' decision will help in delinating homogeneous agricultural situations. Extension, ZRS Researchers and NGOs take part in delination of homogeneous niches.

5.21 *Farmers' Situation Analysis and Problem Formulation.* Representative sites from different homogeneous agricultural situations are selected. PRA tools are employed to identify production problems, socio-economic constraints and opportunities confronting different types of rural households. Farmers, extensionists, researchers and NGOs participate in the analysis.

Prioritization

Problem Prioritization

5.22 The first-three steps form the diagnostic phase. Its outcome is a better appreciation of farming situations and identification of a large number of problems. The next step is prioritizing the problems. A participatory problem prioritization based on the criteria of regional annual productivity losses helps in ranking problems that inflict largest losses. Effective problem solving will ensure greatest incremental contribution to research/extension efforts. Prioritization is best carried out through active participation of farmer, extensionist, researchers and NGOs.

5.23 At this stage, problems that have technological solutions are transferred to the extension system for programming, while the remaining problems that await technological solution are picked for subsequent phase of activities.

Planning and Designing

Planning and Designing Technological Solutions

5.24 This step involves linking formal and informal R&D to develop possible solutions to farmers' problems, given their needs, opportunities and institutional environment. Researchers consult farmers, extensionists and NGOs in this stage.

Implementation

On-Farm Research

5.25 Depending on the nature of prioritized problems, studies can be initiated either at the experiment station or in the farmers' field. While experimental station research is Researcher-designed and Researcher-implemented, the recommendation formulation await findings of on-farm research and technology validation.

5.26 Four generic types of farmer participation in implementing on-farm research, viz. contractual, consultative, collaborative and collegial modes of implementation (Biggs, 1989) are described below.

Contractual

5.27 Scientists/Extensionists enter into contract with farmers. Farmer provides land or services. Professionals make little or no effort to seek her or his views. Professional designed and professional implemented experiments such as experiment on cultivators' field (ECF) is a good example of contractual participation. Farmers' role is passive and participation is not an explicit objective.

Consultative

5.28 Scientists/Extensionists consult farmers about their problems and then develop the solutions. It is professional designed and professional-implemented trials. Farmers' inputs are actively sought at diagnostic phase to define farming system and diagnose priority problems and subsequently in the evaluation of proposed solutions. Professionals design experiments to test various solutions or to better understand identified problems. Consultative mode has been predominantly used in on-farm research with farming systems perspective.

Collaborative

5.29 Scientists/Extensionists and farmers collaborate as partners in the research process. Professionals actively draw on farmers' knowledge and experimentation in solving problems confronting the farming community. It is Professional/Farmer-designed and Farmer-implemented experimentation. Farmer contributes actively both in ideation and implementation at every stage of technology development and dissemination.

Collegial

5.30 Scientists/Extensionists strive to strengthen farmers' informal research and development system in rural areas. It could take two forms: encouraging Farmer-designed and Farmer-implemented experiments for prioritized problems or, Farmer-designed and Professional-implemented experiments to unravel the underlying logic of the indigenous knowledge system. The emphasis is placed on enhancing the ability of farmers to carry out their experimentation systematically, as well as to request information and services from the research and extension system.

Multi-Location Testing

5.31 The on-farm research for technology generation is restricted to a few in number. Based on its results, large scale multi-location trials are designed for fine tuning of technology. Fine tuned technologies, in turn, are transformed into extension messages. Multi-location trials are preferably implemented in collaborative mode.

5.32 In both phases of implementation, extensionists, besides selecting farmers, help researchers manage the trials. Researchers ensure scientific rigor, organize and process data from the trials. Researchers, extensionists, farmers may all have valuable insights to contribute to the interpretation of trial results.

Evaluation

Evaluation and Feedback

5.33 At different stages of on-farm research/extension, farmers–both collaborators and non-participants, but members of farmers' organization – are provided opportunities to evaluate the experimentation employing their own criteria. Researchers not only elicit farmers evaluation criteria but also incorporate the same in their subsequent experimental strategies. The participatory approaches described here, also provide farmers a mechanism of expressing their needs to research, extension and policy makers through the diagnostic phase. Thereby, this methodology strengthens the feed forward and feed-back from farmers to the public system.

Dissemination

5.34 Dissemination of technologies is identified at two phases during the sequential activities. After prioritization, problems that are well understood and have technological solutions are taken by the extensionists for adaptive research. The fine tuned technology is then transformed into extension messages.

5.35 Subsequently, technological recommendations emanating from the farmer participatory approaches are transformed into different types of extension messages. Some of these are designed to inform farmers of the new recommendations, others aim to motivate farmers to apply the recommendations. Still others instruct farmers on how to put into practice the recommendations.

5.36 Farmer-to-farmer diffusion of effective and economically viable technologies takes place as an autonomous process. Monitoring such horizontal diffusion pattern and farmers' adaptation behaviour would also assist in formulation of messages that transforms recommendations into farmer behaviour.

5.37 The emphasis in the methodology is on designing means of enhancing farmers' control over the agenda for research and extension and on improving the feedback process. However, farmers will demand only those options which they have experienced. Farmers may not be aware of a range of technological opportunities and hence they may never demand them. But such newer opportunities emerging due to scientific breakthroughs lie within the knowledge systems accessible to scientists. The major challenge confronting research and extension professionals is that of forging a judicious blend of science-led and bottom-up approaches and thereby bringing together the best each side has to offer.

5.38 One of the several ways of bringing supply-pushed and demand-pulled approaches of agricultural extension and research is to exercise greater care in selecting entry point strategy. The professionals in order to establish credibility, select a few of the prioritized problems that have a greater probability of being successfully solved within the socio-economic constraints of the farmers. Subsequent sequencing strategies, can then allow researchers not only to be sensitive to complex problems but also create space for exploring newer opportunities that are science-led and thereby improve goodness-of-fit with biophysical farming situation and enhance the livelihood sustainability of the resource poor farmers.

5.39 A select group of NGOs have distinguished in innovating participatory methods of diagnostic and farming systems, approach to agricultural technology development; innovations in technologies and management practices; dissemination methods, and improved training methods. The NGOs have also experimented with institution building and human resources development aspects of extension.

5.40 Analyses in the preceding chapter also revealed that weaknesses of public agricultural extension and research systems, particularly in the areas of institution building, human resources development, articulation of and sensitivity to the needs of disadvantaged groups of farming community, obtaining farmers' evaluation and feedback can effectively be bridged through NGO collaboration. Similarly, some of the innovative NGOs are well placed in providing experiential learning to the staff of public systems in participatory methodology of diagnoses, agricultural technology development and farming systems research/extension, gender-variable methodologies and sustainable agriculture.

5.41 In addition to the above roles, NGOs and public agricultural research systems could play a wide range of roles in interacting with each other in agricultural technology development. Based on a series of case studies of relationships and linkages between NGOs and NARS, Farrington *et al.* (in press) have identified four sets of roles:

(i) NGOs disseminate the results of research conducted by NARS;
(ii) NGOs and NARS conduct research jointly, and NGOs disseminate;
(iii) NARS disseminate the results of research conducted by NGOs; and
(iv) NGOs disseminate the results of their own research.

Management of Farmer Participatory Process

5.42 Table 5.1 contrasts the existing public agricultural extension and

TABLE 5.1: **Emerging Features of Public Agricultural Extension and Research Systems**

Features	Existing	Emerging
1. Source of Innovation	Centrist	Multiple
2. Technology development	a. Top-down Science-led	a. Balance between bottom up and science-led
	b. Experiment station based	b. Complementary relationship between on-station and on-farm research
	c. Component technologies	c. Farming systems perspective
3. Technology Dissemination	a. Uniform structure	a. Increasing role for Farmer Organization in the structure
	b. Non-differentiated dissemination process	b. Agricultural situation specific participatory process
	c. Technocratic	c. Technology transfer blended with institution building and human resources development
	d. Crop-based	d. Broad-based, farming systems perspective
4. Clients	Male-oriented	a. Gender-differentiated
		b. Enlarged coverage to rural youth and agricultural landless labour
5. Interface with farming community	Selected 'contract' farmers	Group approach farm extension volunteer provide interface between external agencies and local community
6. Programme Planning	a. Top-down, centralized	a. Bottom-up, decentralized
	b. Rigid and fixed	b. Flexible, participatory programme planning
	c. Non-responsive	c. Responsive to local situations
7. Responsibilities of VEWs	Extension and non-extension	a. Locally determined agenda

TABLE 5.1 (*Contd.*)

Features	Existing	Emerging
	determined top-down	b. Back up technical support
		c. Responsibility matched with authority
8. Monitoring and Evaluation	a. Routine	a. Creative, output/impact oriented
	b. Monitoring inputs	b. Monitor adaptation, adoption and rejection
	c. No mid-term	c. Mid-term corrections
		d. Monitor horizontal diffusion
9. Incentives and Rewards	Absent	Performance-linked
10. Cost Structure	High	Cost reduction strategies

research systems with the emerging roles that they should embrace in order to become farmer participatory. Institutional changes envisioned in making public extension and research systems farmer-participatory and client-accountable should be translated into appropriate management systems. Three major areas of extension management have been identified to foster farmer-participatory process. They are decision making, communication support and management control.

Decision Making

5.43 The present extension programme planning and decision making is centralized. It does not provide manoeuvrability at the local level. Centralized decision-making should be substituted with decentralized programme planning.

5.44 The diagnosis exercise results in outcomes that can be classified into researchable problems, extension needs and developmental needs. The VEWs agenda is formulated locally based on the villagers' emerging needs. Both farmer organizations and farm extension volunteers (FEV) play a critical role in decentralized extension programme planning. Participation of representatives from extension, research, NGOs and farmer organization in the programme planning will contribute significantly to its success and legitimacy. The decentralization of programme planning control also allows priorities to vary as per the local needs. The VEW feeds forward his agenda based on local extension needs and agricultural development requirements and to the AEO and District level for negotiating back-up support activities and thereby attaining his extension objectives.

Back-up Support

5.45 In the existing system a routinized technical training is provided to the VEW. However, that should be replaced by a need based back-up support. VEW may need newer skills in need identification through PRA, project planning, group formation and improved communication.

5.46 However, the success of decentralized extension programme planning hinges on adequate technical and financial support to the VEW for implementing locally decided activities. The farmers organization demands services from the VEW, who in turn articulates the local needs in the form of specific projects upwards. In bottom up planning, decentralized needs should influence the central or State-sponsored special schemes also. The Subject Matter Specialists will, henceforth, design training programmes based on training needs identified through this process. The process is by design demand-driven. Responsibility is matched with decentralized decision-making authority.

Monitoring and Evaluation

5.47 In addition to the hierarchical monitoring, farmer organization is also involved in monitoring and evaluation of local extension staff. Thus, the current monitoring indicators, such as "number of visits made to farms", "number of meetings arranged", and the "number of demonstrations/minikit trials organized" may change to "activities initiated in response to local level needs", "resource mobilization for problem solving", "impact per person", "impact per unit area", or "effectiveness in solving some of the prioritized problems". Besides horizontal diffusion, farmers adaptation, adoption and rejection behaviours will be concurrently evaluated for mid-term programme corrections.

Incentives and Rewards

5.48 Though the M&E units regularly monitor the VEWs' performance, it is not linked to any incentive or reward system. Nor have many States considered seriously the need for planning career advancement of village extension worker. Morale of local level extension staff is waning.

5.49 Decentralization of decision making and flexibility in programming should be reinforced through instituting appropriate performance-linked incentives and reward system.

5.50 It is proposed that a graded system of incentives and rewards be instituted for the VEW. Farmer organization's evaluation of performance is given due weightage in designing a graded system of reward that reinforces VEWs farmer participatory behaviour and the impact thereof. Under the scheme all effective and farmer-responsive VEWs would receive citations of merit; about 10 per cent of such VEWs who have been high performers will qualify for one time monetary incentive of Rs. 500 annually; while the best VEW in the block will receive an award and a cash prize of Rs. 1000.

Flexible Funding for Innovative Approaches

5.51 In most of the DOAs, untied funds are not available to support innovative experiments at the local level. In Rajasthan, DOA has a scheme of untied funds for innovative programmes. Under this scheme, the Secretary, Agriculture Production can sanction, without reference to Finance or Planning Departments, any small innovative scheme or programme received from field officers for field testing over one season or one year. However, these funds cannot be utilised for purchase of vehicle, sanction of staff, acquisition of equipment, construction of buildings etc. Under another scheme knowns as "BHAGIRATH", any VEW can propose innovative ideas or set any extraordinary target for himself in any agricultural activity or programme for implementation at the local level. The DOA not only provides technical and financial support but also assures that the field level functionary will not be shifted during the tenure of experimentation.

5.52 It is suggested that the State DOAs should earmark atleast 5 per cent of their annual plan budget for flexible funding of innovative ideas that emerge from the field level. This untied fund can be released to VEW's for implementation of innovative programmes. A screening committee may be constituted at the district level to screen innovative proposals and recommend short listed innovative ideas for award.

5.53 All Central Sector and Centrally sponsored special programmes should also set aside 10 per cent of their outlays for supporting farmer participatory approaches. The fund may be utilized for the following purposes that subserve programmatic thrust.

(i) Horizontal diffusion	Farmer exposure visits, Innovative farmers' workshop
(ii) Farmer-participatory	On-farm research and adaptive trials experimentation for solving local problems.
(iii) Technology dissemination	Use of mass media, literature, field days, farmers fair, visit to experiment station, etc.
(iv) Human resources development	Groups formation and servicing farmer organisations.

National Programme on Farmer Participatory Extension and Research

5.54 In order to encourage greater farmer participation in agricultural technology development and dissemination, it is recommended that a new National Programme be launched. This will create an enabling environment for State line departments, NGOs and Agricultural Research Institutions to reorient their research and extension and make them more accountable and responsive to the farmers' problems. The programme will provide grants on a competitive basis; State departments, NGOs and Zonal Research Stations can bid proposals for funding.

Research Programme Planning at Zonal Research Station

5.55 It is proposed that the ZRS be indirectly made accountable to the farming community. At present, research at the ZRS is sponsored either through ICAR's, AICRP or State funds. The ZRS researchers should advocate modifications of respective on-going AICRP project in the light of regional prioritized problems that emerge from farmer participatory process. Of course, researcher should also undertake anticipatory problem solving research at the ZRS that may never be articulated by farmers.

Contractual Research

5.56 The State research resources should be allocated exclusively for farmer participatory research and adaptive research on a contractual basis. The prioritized researchable problems are consolidated at the district and region by the DOA in close collaboration with the concerned ZRS researchers. This forms the basis for research resource allocation at the State level.

5.57 Other line departments, State Seed Corporation, Agro-Industries should have a lumpsum provision for contracting research to ZRS on relevant problems. The contractual research could take the form of a commissioned study or research award on proposals by ZRS researcher.

Joint Extension – Research Management Board

5.58 It is proposed that to oversee the farmer participatory processes of extension and research a Joint Extension—Research Management Board be constituted at the Zonal level. It would have respresentation from research, extension, NGOs, farmers organization and agro-input dealers.

Representatives from farmers organization would be on the basis of homogeneous farming situations/recommendation domain. Each farming situation in an agro-climatic zone would be represented by innovative farmers nominated by the farmer organizations. The upper limit of the Joint Extension – Research Management Board will be 18, with at least two representatives from the agro-input dealers. The Associate Director of Research/Extension of the respective ZRS will be the convener. Its functions will include:

1. To oversee the Zonal research and extension programme planning;
2. To review and allocate resources to zonal extension and research activities;
3. To monitor the programme implementation; and
4. To mobilize resources for execution of farmer-relevant research and extension programmes.

5.59 The Joint Extension Research Management Board will meet once in a month and will have executive responsibilities. It is in addition to the Zonal Research Extension Advisory Committee.

5.60 It has been observed that research programming suffers from weak management at the ZRS. In addition to contractual research and competitive bidding for research grants, the following procedures may streamline the research programming and the research managers' ability to monitor important milestones.

(i) Detailed specification of the intended outputs from a research project;
(ii) Outline of important milestones and adherence to a programme of work to deliver those outputs;
(iii) Strengthening research managers' capability to monitor whether such a programme of work is on schedule;
(iv) Identification by researchers of who the users of research output will be; and
(v) Resource mobilization from corporate end users.

Subsequent rounds of funding will be contingent on researchers' proving effective in delivering goods.

5.61 It must be mentioned here that changes in institutional values and rewards in the research system appear to be complex. Nevertheless, an attempt must be made to make the research system responsive to the clientele. Though, it would not be fair to design a performance-linked reward system for extensionists and not for researchers.

Improving Communication Support

5.62 Radio and electronic media can be effective and efficient means of disseminating certain types of technical and marketing information and of creating farmer awareness of new technology. Efficiency and effectiveness of field staff in group meetings and discussion, adaptive research/demonstrations and farmer field days can be greatly enhanced through proper use of teaching aids, audio-visual and other mass media material and extension publications.

5.63 A clearing house mechanism is also required at national level for processing, indexing, abstracting and preparation of reader-friendly documents on innovative extension strategies, approaches and methods and extension management for extension staff at different levels in the public, private and non-government sector.

5.64 It has been observed that currently no distinction is made between 'technology', 'recommendation', and 'extension message'. Though information distortion is recognized as technology is passed down from research to extension to farmers, yet little attention is being paid to information transformation as a deliberate activity. For instance, a systematic practice has not been evolved for translating a technical recommendation, say 60 kg per hectare of P_2O_5, into farmer behaviour or behavioural objectives. Such a translation of recommendation into behaviour is essential for adequate farmer adaptive testing.

5.65 Similarly a related problem arises due to lack of distinction being made in current practice between the recommendation and extension messages. Yet, in actual practice, one recommendation can lead to many different messages. Some of them might seek to inform farmers of the new recommendations, others try to motivate farmers to apply the recommendation, still others instruct farmers in how to apply. The absence of clarity in the concepts used leads to sloppy procedures, inefficient media planning, flawed targetting and blindness to mistakes.

5.66 Market intelligence is another area which has not received adequate communication support in India. However, with the recent thrust on agricultural export zone and trend towards commercialized agriculture, market research and intelligence would be of permanent importance for farmers to respond to dynamic market forces and adjust their production decisions.

5.67 It is, therefore, suggested that at each State an extension communication unit be established and staffed with communication, behavioural science and marketing professionals. Besides evolving appropriate information transformation process, the State level extension communication unit will be responsible for media planning, and designing effective and efficient means of disseminating technology message, extension message and marketing information. The unit will also experiment with innovations in generating feedback from farmers and field-level functionaries for responsive programme planning. Developing and supporting such extension communication unit at the State levels should be viewed as an essential component of an effective farmer participatory agricultural extension system.

Educational Qualifications and Training Needs

5.68 The study team found that the educational qualifications of extension personnel, for historical reasons, have remained very low in relation to the assignments and responsibilities they are expected to carry out. The monthly and fortnightly training workshop is too narrowly focussed on crop technologies.

5.69 A substantial training effort is needed to upgrade the technical and managerial/organizational skills, knowledge and qualifications of field and junior extension personnel. Some of the emerging areas of training are: Gender-related training, sustainable agriculture, participatory rural appraisal, participatory development, systems management, mobilisation of groups, small group dynamics, etc. It may be noted here that the capacity of Zonal Research Stations and State Agricultural Universities, which supply master trainers, is limited. Innovative and action-oriented NGOs may have comparative advantage in providing medium-term and short-term trraining. In the long-term, institutions like MANAGE should take the lead in building a cadre of master trainers for fulfilling the training needs. Simultaneously, substantial assistance is required in strengthening both the human resource capacity of extension to discharge the role of human resource development and technology transfer, and the State Agricultural Universities that train extension personnel.

Cost Reduction Strategies

5.70 Currently, public sector extension, private sector firms, NGOs, Cooperatives and semi-autonomous organizations are discharging the role of technology transfer in India. In certain commodities/regions/ sectors, technology has already shifted from being a "public good" to becoming a "private good". In this process, technology transfer costs are mostly paid by the farmers themselves, either through price of purchased inputs or through contracted consultancy services. In our classification scheme, high value commercial agriculture region is ripe for privatization. Similarly, a series of plant protection services in assured irrigated and high value crops in rainfed area is in the process of being contracted out.

5.71 A publicly supported extension system is however, essential to improve productivity, increase output and enhance food security in the majority of rainfed regions. In addition, along wih concerned NGOs, public supported extension system should begin addressing issues of environmental degradation, and management of natural resources associated with sustainable agriculture.

5.72 Recommendation on extension volunteers to be employed by farmers' organisation is likely to reduce budgetary burden on the exchequer. Another way for resource poor farmers to lessen the unit cost of extension services and for the government to increase its level of extension coverage is for extension to assume institution building role. It assists farmers to get organized into functional farmer organisations.

This will accelerate farmer-to-farmer diffusion of new technology, increase accountability and responsiveness of extension and research, and increase the efficiency of extension in reaching a larger majority of its intended clientele.

5.73 Experimentation on cost-reducing extension strategies also needs to be designed and implemented through the farmers' organization. The farmers' organization can identify and hire, from educated unemployed youth, potential extension agents, under mutually negotiated terms and conditions of employment and establish norms of performance evaluation and rewards. Extension, on the other hand, can cover a fixed proportion of total costs which taper off once the experimentation is found to be viable and self-sustaining. More important role for extension, however, would be that of 'holding hand' operation and providing back stop support in terms of creating opportunities for experiential learning, on the job training, publications, audio visual led distance learning and related support that is amenable for a centralized operation because of economies of scale. Also extension would foster networking relationship amongst such locally appointed extension volunteers as well as between the extension volunteers and zonal research station researchers.

Policy Support for Farmer Participatory Extension and Research

5.74 Continuation of reform processes to make the public extension and research systems farmer-participatory and client-accountable would call for a series of policy support. Some of the illustrative enabling policies are:

(i) Formulation of Agricultural extension and research policy with a predominant role for farmer in future extension and research programme planning.
(ii) Policy concurrence for transformed role of extension.
(iii) Acceptance of the concepts of Farm Extension Volunteers and Farmer Organization as measures of increasing involvement of local community, expanding extension coverage and as a cost reducing strategy.
(iv) Policy approval of concepts of flexible funding to promote innovative approaches, contractual research and performance - linked incentives and awards.
(v) Policy decision on broadbasing of extension imply

identifying 2 or 3 major farmers' land-based needs in a specific zone that can then receive focussed attention. The pace, extent and modalities of broadbasing should be decided at the State level.

(vi) Policy decision in the rainfed regions to declare dominant enterprise department as the lead department. Consequently all field extension personnel of DOA would be placed under the administrative control of the lead department.

(vii) Policy concurrence on constituting Joint Extension-Research Management Board at the agro-climatic zone level for coordination, overseeing and resource allocation of extension and research programmes.

II. AGRICULTURAL SITUATION SPECIFIC RECOMMENDATIONS

5.75 In the earlier section, we outlined recommendations in terms of institutional reforms that must be continued in the T&V System and Zonal Research Station. In the present section, multiple options for different agricultural situations are developed and evaluated. The Tables 5.2 and 5.3 are suggestive of distinctive features of extension and research approaches respectively.

II.1 Options for Irrigated Areas

Agricultural Research

5.76 The farmer participatory methodology outlined in 5.18 will enable the ZRS to modify its research agenda. A multi-disciplinary team of researchers at ZRS collaborates with extension SMS, SAU's district level scientists[1], block level extension team[2] and NGOs, if available, in identifying both short and long term problems.

Since sustainability issues have become critical in this region, the researcher team will design a protocol for farmer monitoring of sustainability issues. This procedure will be mounted on representative farmers plot and monitored by farmers. Extensionists will assist in supervising the programme and creating data base on indicators of non-sustainability.

5.77 The farmer monitoring will enable researchers to initiate both anticipatory research (OSR) and farmer participatory problem solving research (OFR) that would complement experiment station research. Several sustainability problems like, drawdown of acquifer, dynamic weed regimes, pathogen build up etc., demonstrate common pool resource characteristics. Moreover, such situations cannot be simulated at experiment station. Therefore, action research mode may be preferred. Groups may be formed around problems among interdependent farmers for evolving effective coping strategies.

Agricultural Extension

5.78 Keeping in mind the existing and newer sets of skills required by farmers and extensionists, we have developed six options for the irrigated situation. Table 5.4 (i) through (iv) portrays options, composition and role of extension team, roles of different functionaries, training needs and evaluation of different options. The options are :

Option 1: Forming an extension team of 2 AEOs and 1 SMS (redeployed from the sub-division) at the block level. The team will have technology transfer, facilitate horizontal diffusion, organize innovative farmers' workshop, attend crop clinic at block, organise exposure visit, kisan mela, field days, etc. enable VEW in group formation, HRD and assist researcher team to conduct on farm research.

Option 2: Forming an extension team of 2 AEOs and 1 SMS (redeployed from the sub-division) at the block level. The team will offer diagnostic and plant and soil care services on demand.

Option 3: Forming a team of SAU and Scientists (4-7) and 1 SMS (redeployed from the district) at the district level. The team will carry diagnostic studies and conduct farmer participatory research at representative sites. It is expected that the SMS – the weakest link in the T&V chain – will have ample opportunity to build competence by collaborating with the SAU Scientists.

Option 4 : Forming a team of SAU Scientists (district) and 2 AEOs plus 1 SMS (block). The research team, in addition to its normal technology transfer responsibilities, will guide and collaborate with the block level extension team. The extension team will perform all responsibilities that have been outlined for option 1, and

Option 5: Forming a team of SAU Scientists (district) and 2 AEOs, 1 SMS and 10 VEWs. The only difference between option 4 and 5 is that VEWs who were redeployed to do development work in the former have been retained in the latter. Consequently, this option provides for

TABLE 5.2: **Salient Features of the Distinctive Extension Approaches for the Different Agricultural Situations in the Country**

Salient Features	Assured Irrigation System	Rainfed farming System	High Value Commercial Crop System
1. Technology Requirement	Specialised	Diverse	Highly Specialised
2. Extension Approach (Dexter 1986)	Promotional (Agency decides on basis of views and needs of farmers)	Participative (Partnership between Agency & farmers at all stages)	Consultancy (Advice on request)
3. Focus on Extension	Crop Based System	Farming System Based	Crop/Tree Based System
4. Programme Content	Standardised Messages	Flexible Messages	Standardised Messages
5. Services	Production Based	Resource Based	Integrated Services (Post Harvest)
6. *CLIENTELE*			
– Extent to which has taken initiative in Participatory Approaches	Desirable	Absolutely Necessary	Not required
– Extension Volunteers	Not Required	Necessary	Not Required

TABLE 5.2 *(Contd.)*

Salient Features	Assured Irrigation System	Rainfed farming System	High Value Commercial Crop System
– Contact Farmers	Nct required	Not required	Not required
– Gender Differentiation	Low	High	Low
7. Role of Extension Personnel	Advisory	Advisory, Enabling, Convening, Consultancy and Tour Operator (Robert Chambers)	Advisory and Consultancy
– Deployment	At Block Level (As a team)	At Village Level	In Production Areas
8. Training Needs of Extension worker	Technological and Problem Diagnosis	Technological and Behavioural	Low Training Needs
9. Media Support	Modern Audio Visual and Print Media Suppcrt	Low Cost Indigenous Audio-Visual Aids	Sophisticated Communication Technology
10. Role of NGOs	Negligible	Important Role in (a) Creating Farmers' Organization (b) Training of Farmers (c) Training of Extension Volunteers (d) Training of Extension workers, especially in Behavioural Sciences and Group Dynamics	Negligible
11. Çost-Sharing	Desirable	To be experimented	Must

TABLE 5.3 : **Distinctive Features of Proposed Agricultural Research in Different Agricultural Situations**

Distinctive Features	Assured Irrigated Area	Rainfed Area	High Value Commercial Agriculture
1. Nature of Problems	* Declining yield	* Low productivity	* Weather-induced losses
	* Sustainability issues, nutrient mining, pathogen buildup, dynamic weed regime, draw-down of acquifer	* Fragile resource base * High risk * Complex interactions among enterprises	* Insect, pest incidence * Management problems * Quality Control
	* Inefficient resource use * Inefficient cropping pattern	* High run-off losses * Problem soils	
2. Methodological issues			
(i) Desirable characteristic of research process	* Farmer monitoring of sustainability issues	* Farmer-centered retrospective diagnostics	* Contractual research through growers' associations
	* Data base on indicators of non-sustainability	* Data base on indicators of non-sustainability	* Problem solving
	* Exploratory research on representative sites	* Inventory of indigenous knowledge systems	* Exploring new opportunities

TABLE 5.3 *(Contd.)*

Distinctive Features	Assured Irrigated Area	Rainfed Area	High Value Commercial Agriculture
	* Long-term experiments at research stations	* Linking of formal and informal R & D	
	* Interactions among inputs and soil-plant relationship	* Discerning farmers' risk-coping strategies - both long-term and short-term	
	* Simulation exercises Commodity-based Systems research	* Farmer participatory approaches to technology generation and dissemination	
	* Research on common pool resource and governance	* Resource-based systems research	
	* Explore new opportunities	* Research on Equitable use of common pool resources	
		* Explore new opportunities	
(ii) Who designs the experiments?	Researcher (increasingly researcher) would have to incorporate farmers' understanding of sustainability)	Participatory Farmers' researcher, extensionists as equal partners	Commissioned Researcher designed on grower identified problems

TABLE 5.3 *(Contd.)*

Distinctive Features	Assured Irrigated Area	Rainfed Area	High Value Commercial Agriculture
(iii) Who implements the experiments?	Researcher+farmer	Depending on nature of problem Farmer alone Farmer and extensionist Farmer, researcher and extensionist	Researcher
(iv) Monitoring and Evaluation	Researcher+farmer	Farmers' evaluation criteria incorporated by researcher and extensionist	Researcher and grower
(v) Methodological Development	Researcher-led blending indigenous knowledge systems with formal methodology	Collaborative emphasis on blending indigenous knowledge systems with formal methodology	Researcher-led
(vi) Interaction between component and system research	Interactive/conflictive interactions, systems research feed forward ideas for, controlled component research	Feed forward and feed back for conducting on-farm and on-station research	Stations based research geared to address farmers' problems

TABLE 5.3 *(Contd.)*

Distinctive Features	Assured Irrigated Area	Rainfed Area	High Value Commercial Agriculture
3. Institutional issues			
(i) Role of public sector	Medium to High	High	Low
(ii) Role of extension Professionals	* Help identify problems * Help mounting farmer monitoring of sustainability issues * Help creating Data base trials and dissemination	* Help conducting diagnostic studies * Help delination of homogeneous agricultural situations * Help identify representative sites * Help managing on-farm research * Help distilling lessons and feedback	Promotional

TABLE 5.3 *(Contd.)*

Distinctive Features	Assured Irrigated Area	Rainfed Area	High Value Commercial Agriculture
		* Demonstrations and dissemination * Human resources development and institution building	
(iii) Role of NGOs	* Minimal * Potential collaborative role in governance of common pool resources institutions	* Important complementary role in methodology development, methodology dissemination, building farmers' organisation, helping to articulate farmers' demands on research services	* Minimal

TABLE 5.4: **Irrigated Areas**

(i) Options and Compositions of Team

2 AEOs + 1 SMS	2 AEOs + 1 SMS	4-7 SAU Scientists + Distt. SMS	SAU Scientist + 2 AEOs + 1 SMS	SAU Scientist + 2 AEOs + SMS + VEW
I	II	III	IV	V
Base = Block	Base = Block	Base = District	Base = District for Researcher. But operation at block level	Base = District for Researcher. But operation at block level
TEAM Farm 1. TOT F.O. 2. Strategy- F F 3. 2 times in season-workshop. 4. Clinic once in a fortnight at block + visit on call. 5. Organise at block level, kisan mela, exposure visit, field days, visit to research station, adaptive trials with F.O. once in a season. 6. Facilitate VEW in IB, HRD, GM. 7. OFR, AT, Field days	1. Clinic concept and visit farmers' fields on call, (enlarge version of S.No. 4 under option 1)	1. Field visit. 2. OFR, Adoptive trial, group meetings, field days, correspondence with farmers, radio talks, newspaper articles	1. TOT Distt. (Res.) Plus All Activities of option 1 at block level	1. IB, HRD, GM 2. All at block (same as clinic) date and time fixed for consultation).

Note TOT = Transfer of Technology; FO=Farmer Organisation; F=Farmer; IB=Institutional Building; HRD=Human Resource Development; GM=Group Mobilization, OFR=On Farm Research; Res=Research; AEO=Agriculture Extension Officer; SMS=Subject Matter Specialist; VEW=Village Extension Worker; AIC=Agriculture Information Centre

TABLE 5.4. (Contd.)

(ii) Roles

Option	AEO	VEW	SMS (Sub-division)	SMS (Distt.)	Researcher	Farmer Group	Farmer
I.	2 in the team at block remaining deployed for development work.	IB-HRD-GM Team FO	Withdrawn deployed at block level.	Support to AEO Team – Information – Occasional participation in meetings – Fortnightly interaction	Guidance for adaptive trial.	Active role	Can seek advice
II.	2 AEOs as Crop Doctors, Remaining deployed for development work.	Withdrawn deployed for development work.	Withdrawn deployed at block level.	Support to AEO Team – Information – Occasional participation in meetings – Fortnightly interaction	Guidance to team.	Active demand role.	Active demand role.

TABLE 5.4 *(Contd.)*

Option	AEO	VEW	SMS (Sub-division)	SMS (Distt.)	Researcher	Farmer Group	Farmer
III.	Development worker and facilitator for field activities	Assist the distt. team in AT & OFR.	Supervise the trial & OFR and assist team. Withdrawn	SMS team withdrawn from conventional responsibilities	—IB, HRD, GM. —Specialised training for farmer group as per need. – Training (on joli) for SMS.	Active role	Can seek advice
IV.	2 AEOs in team, others deployed for work.	All deployed for development.	Withdrawn and deployed in team.	Withdrawn	– Back up technical support. – Facilitate IB, HRD, GM.	Active role	Can seek advice.
V.	2 AEOs in the team IB, HRD, GM Others deployed for development	10 VEWs with team and rest withdrawn for development work.	– Withdrawn – Deployed in team.	Withdrawn	– Back up technical support. – Facilitate IB, HRD, G.M.	Active role	Active role

Note: AEO=Agriculture Extension Officer; VEW=Village Extension Workers; SMS=Subject Matter Specialist; IB=Institutional Building; HRD=Human Resources Development; GM=Group Mobilization; FO=Farmer Organisation; AT=Adaptive Trial; OFR=On Farm Research

TABLE 5.4 *(Contd.)*

(iii) Training Needs

Option	Functionaries	New skills	
		Behavioural Science	Agricultural Science
I.	AEO, VEW, SMS	IB, HRD, GM at Zonal Level.	– Technology and Diagnosis. – Participatory Research Methodology
II.	AEO, SMS	Diagnostic skill + technology	
III.	Researcher	IB, HRD, GM	Technology, Participatory Research Methodology
IV.	AEO + SMS	IB, HRD, GM	– Technology, Diagnostic + Participatory Research
	Researcher	IB, HRD, GM	– Participatory Research Methodology
V.	AEO, SMS	IB, HRD, GM	– Technology Training, Diagnostic Skills
	Researcher 2 AEOs in team	IB, HRD, GM All deployed	– Participatory Research Methodology

Note:
AEO=Agriculture Extension Officer;
VEW=Village Extension Worker;
SMS=Subject Matter Specialist;
NGO=Non Governmental Organization;
IB=Institutional Building;
HRD=Human Resource Development;
GM=Group Mobilization

intensive group mobilization and HRD also.

5.79 The roles and training needs of extension is detailed in Table

TABLE 5.4 *(Contd.)*

(iv) Evaluation Criteria

Option	Professional competence building	Coverage	Farmer participation	Responsiveness	Political Compatability	Administrative Feasibility	Accountability	Cost Effectiveness
I	I	III	II	II	III	III	II	III
II	I	I	I	II	I	II	II	II
III	II	I	III	II	I	I	II	II
IV	III	II	III	III	I	I	III	III
V	III	III	III	III	III	III	III	II

Note:
I=Low Weightage; II=Medium Weightage; III=High Weightage

5.4 (ii and iii). We have developed a set of evaluation criteria that includes:

1. *Professional Competence Building*: In irrigated areas, the extensionists have not been able to keep pace with the farmers' knowledge and skill on science led agriculture. Therefore, the preferred option should create opportunities for extensionists to enhance their professional competence.
2. *Coverage*: The public extension coverage is not wide. The short listed option should provide for a greater coverage of population and area.
3. *Farmer Participation*: The selected option should envision greater depth of participation. The quality and number of farmers covered both are important.
4. *Responsiveness*: Responsiveness denotes ability of the system to respond to the real needs of farming community and respond fast enough to the dynamic socio-technical environment surrounding local farming systems.
5. *Political Compatibility*: It reflects the likely ease in carrying

out the reform even in the face of opposition from those segments of staff who would be losers. Political compatibility should be high for an option to be preferred i.e. those who tend to lose out should be outstripped by the gains from the option.

6. *Administrative Feasibility*: Administrative ease in reorganizing the public extension and research is denoted by administrative feasibility criteria.
7. *Accountability*: Accountability addresses the issues of effectiveness of staff as well as effectiveness of programme in solving clients' problems.
8. *Cost-effectiveness*: The preferred option should trigger, for a given cost, other things remaining the same, a greater stream of benefits.

5.80 The study team's evaluation of the five options is shown in Table 5.4-(iv). The preferred option was the fifth one. Option 5 provides for functional integration rather than territorial integration of research and extension system, which should not pose administrative problems. The specific tasks are :

1. Dissemination technology, group mobilisation, institutional building and human resource development.
2. The programme of field visit will be drawn by the team as per need of farmers. The team will ensure their interaction with village level farmers groups/organizations in each homogeneous agriculture situation on rotational basis.
3. Team shall be entrusted with the tasks of conducting on farm research, adaptive trials, field days on innovative farmers fields, exposure visits of farmers to research station and other areas.
4. Team of extensionists will develop strategy for communication of information related to technology through mass media for wider coverage.
5. Organise farmers training (village extension volunteers, farm women, youth) in collaboration with KVK/farmers training centres and NGOs to develop diagnostic skills and technical know-how.
6. Team will pass the feedback obtained from the farmers to researchers.

5.81 The farmer participatory methodology will provide the basis for decentralized project planning. The projects will be implemented through the office of Deputy Director of Agriculture at the district level. Sub-Divisional Agriculture Officers will provide logistic support to both scientist and extension team.

5.82 The collaboration will be strengthened through joint professional activities (joint problem diagnosis, joint priority setting and planning exercises, formal collaboration in on-farm research and dissemination activities, joint decision making on release of recommendations, regular joint field visits, informal consultations, joint training activities or seminars and joint publications) and evolving resource allocation procedures through the Joint Extension Research Management Board at the zonal level (formal guidelines for allocating time for collaborative activities, specific allocation of funds and staff secondment for collaborative activities).

II.2 Rainfed Situations

Farming Systems Approach

5.83 The most important change required in the research and extension strategy for the rainfed areas is a shift from crop centred technology generation and dissemination to a farming systems based research and extension approach. As has been brought out earlier, in the rainfed situation, more than in any other situation, the farmer, for a variety of considerations, practices a diverse mix of occupations, in which non-crop enterprises often play a more dominant role and contribute in a larger measure towards income generation than crop husbandry. Unless the research and extension effort takes into account this diversity, it is not likely to succeed in creating an impact.

5.84 The farmer takes a holistic view of his needs and is not governed by the administratively influenced responses of governmental systems. He/she sees his/her crops, livestock, tree farming, aquaculture, off-farm activities etc. as an integrated whole which is inter-related and interdependent. Governmental systems, for administrative convenience, respond in the form of different institutions addressing themselves to different sets of issues like crop farming, tree farming, livestock rearing, credit, marketing, etc... often working in isolation from each other, and, occasionally, even at cross-purposes. Hence all research and extension effort should be remodelled to emphasise a farming systems approach.

5.85 There are several institutional mechanisms by which this reorientation can be brought about. Before, however, evaluating these different options, it would be necessary to point out that there is now a consensus amongst States—a consensus promoted by Government of India about the need for extension to handle all land based activities. In fact, in a National Seminar on T&V organized by Government of India in 1989, this issue was discussed at considerable length and States, while endorsing the need for broad basing of extension, wanted the pace, extent and modalities of this re-orientation to be left to them to decide, in accordance with the prevalent agricultural situation.

Option 1: Enable VEW to become a multi-disciplinary agent

5.86 The first option – infact an outcome of the aforementioned conference – is to make the VEW a multi disciplinary worker in the sense that the SMS from the relevant line departments would provide the requisite skills relating to that department through the fortnightly/monthly workshops under T&V. The motivation behind this option is the acknowledgment in several quarters that it would not be possible, in view of the severe resource constraints being faced by State Governments, for all departments to have their own extension agents upto the field level. Also, it was felt that it would not be desirable for several extension agents approaching an individual famer. It is this model which the government of India has been advocating and some States have taken some initiatives in this direction. This option would be desirable only in those rainfed situations where Agriculture in the narrow sense of Crop Husbandry is the dominant enterprise.

Option 2: Dominant Enterprise Department assumes the lead role, Multi-skilled VEW placed under the lead department

5.87 Where the dominant enterprise in an area is an enterprise other than crop husbandry like Horticulture, Livestock Development, Aquaculture etc. a variant of this option could be explored in which the field extension personnel of the Agricultural Department is placed under the administrative control of the Department which looks after the dominant enterprise.

5.88 This would imply for instance that in some pastoral districts of Western Rajasthan the extension personnel would be placed under the administrative control of the Animal Husbandry Department who would train and employ the service of these personnel for livestock development activities for which there is no network of livestock extension agents at present. Or in some horticulture districts of Himachal Pradesh placing the extension personnel of the Agricultural Department under the administrative control of the Horticulture Department. This institutional mechanism would ensure a single line of command and sharply focus on the dominant agricultural activity of an area. It would of course need re-orientation in the role of extension personnel for which intensive training would require to be organised. An illustration of this organizational structure is found in the Nilgiris district of Tamilnadu where the personnel of the Agricultural Department have been placed under the administrative control of the Horticulture Department as Nilgiris is a predominantly Horticulture District.

5.89 The underlying assumption in this option is that since agriculture in the narrower sense or rather crop husbandry is only a subsidiary occupation and that too subsistence in nature, it makes little sense to have a crop centred approach when the more dominant needs of the farmer are in a different enterprise for which extension support is either non-extension or weak, corollary would be that monthly meetings etc. would be convened by the department whose enterprise is dominant in that area and agriculture department would provide SMS in these workshops. Though this option would require administrative reorganisation it is more likely to work effectively towards promoting the concept of farming systems approach in both extension and research. It is therefore being advocated.

Option 3: Forming a Team of Extension Agents from Different Departments and AEO-Led.

5.90 A third option would be to create a team of extension workers drawn from different departments who would interact as a team with client groups. This team would be coordinated by the AEO of the Agriculture Department. The underlying assumption behind this option is that it would be difficult to create the necessary skills/expertise in the VEW of the Agricultural Department and therefore each department should do its own extension, albeit as a team. The other debatable assumption implicit in this option is that most relevant departments actually have an extension network in the field comparable to the Agriculture department–an assumption which is not quite true. Also there is no attempt at functional integration; it believes that merely by putting extension workers drawn from different departments together in a team (which as has been pointed out earlier, may also not be feasible for want of extension personnel in sufficient numbers in other departments), a farming systems approach or broadbased extension would be introduced. This belief is to say the least quite misplaced which is why the efforts in this direction in Uttar Pradesh in the form of "Kisan Sewa Kendras" or in Assam through the F.M.C.'s is not yielding the desired results.

Option 4: A Corporate Entity

5.91 A fourth option to meet the broadbased needs of farmers is to create an organization which draws on extension personnel of different line departments and brings them under one umbrella. This organization would kind of provide a single window service for farmers. The best example of this approach is the Dryland Development Board of Karnataka which, with the watershed as a unit, has brought into its fold extension personnel drawn from a large number of agriculture and allied sectors who function as an integrated team when interacting with farmers. This option, while feasible when working intensively over small areas like watersheds may not work when replicated over large State systems. This is because it would then imply putting together the major agriculture and allied activities departments as a whole into one giant monolithic organisation which would not be acceptable to any group.

5.92 The options are summarized in Table 5.5-i and their evaluation is shown in Table 5.5-ii. The study team employed the same set of criteria, viz., Professional competence building, coverage, farmer participation, responsiveness, political compatibility, administrative feasibility, accountability and cost effectiveness.

5.93 The option best suited to a farming systems approach is the one being presently advocated by Government of India wherein the VEW is put through periodic training programmes – institutional and peripatetic – for acquisition of skills for dissemination of extension packages in allied subjects with the difference that where the dominant need of the farmer is either pastoral or horticultural or aquaculture based, i.e., the dominant need is in a field other than crop husbandry in such situations the lead department should take on the lead role for broad basing for which the field level extension staff of the agricultural department would be placed under the administrative control of the lead department.

TABLE 5.5: **Rainfed Situations**

(i) Options and Composition

I	II	III	IV
Multi-Disciplinary VEW Provided Training in Broad based extn. by SMS of relevant line depts. But working under Admn. control of Agrl. depts. as in many States	Multi-Disciplinary VEW Provided training in Broad-based extn. by SMS of relevant line depts. But working under Admn. control of lead dept. of the area like in Nilgiris Dist. of Tamilnadu	Team of Extn. workers drawn from Different Deptts. and led by AEO of Agrl. Dept. as in Assam, UP.	Team of Extension workers from Different Depts. but functioning under an umbrella orgn. like Dry-land Dev. Board in Karnataka

(ii) Evaluation Criteria

Evaluation Criteria	Options			
	I	II	III	IV
Professional Competence	**	**	***	***
Coverage	***	***	*	*
Farmer Participation	***	***	**	**
Responsiveness	***	***	*	**
Political Compatibility	**	*	***	*
Administrative Feasibility	**	**	*	*
Accountability	**	**	*	***
Cost Effectiveness	***	***	**	*

Group Approach

5.94 The second most important reorientation required would be to get extension to work with groups than with individual farmers. It is now generally recognized that working with groups than with individual farmers would be more conducive to effective extension.

5.95 The justification for the group approach is based on the following major arguments :

(a) More people can be reached by extension workers.
(b) Groups provide a more effective learning environment.
(c) Many of the proposals made by extension are more appropriate for group action than for individual action.
(d) Groups provide vital feed-back to extension regarding relevance of recommended technologies to field situations.

5.96 Groups enable extension to more effectively overcome the problems of poor farmer coverage inherent in the individual contact approach and assist in the quicker spread of technology.

5.97 Group learning tends to induce better reinforcement of dissemination packages, provides for horizontal diffusion, involves farmers in planning, learning activities, group action etc.

5.98 Further, there are many agricultural operations or extension recommendations which require group mobilization. For instance, integrated pest management, common property resource management, water management, marketing etc. require mobilization of farmers into organized groups for voluntary action for self development.

5.99 Groups tend to provide better feed-back to extension of problems and needs. Additionally, they also facilitate access to services, credit, inputs and marketing.

5.100 This self help approach is based on the assumption that the motivation for rural transformation must come from the people themselves. Outside help must come in response to the expressed needs of the communities and concentrate on fostering local institutions for cooperative self-help.

5.101 An increasing number of countries including India are employing this participatory approach in extension as a way of reaching large number of small farmers effectively and involving them in planning, learning activities, group action and feed-back to extension of problems and needs.

5.102 In fact the objective of achieving broad based agricultural development in which access to research, extension and inputs is more equitable, and in which farmers participate effectively, would require extension not to approach people individually but in groups.

5.103 While it is being recommended that the individual contact farmer approach under T&V be substituted by a group approach, it is felt that given the complexity of the tasks involved in group formation and the inadequate skills of the extension personnel in this regard, priority should be accorded to use any farmers' groups, association or committees already in existence whether or not it is directly involved in agriculture. Farmers form groups or become members of existing ones with a definite objective. There are some factors that pull people together to form groups; groups have developed around credit in Malawi, or marketing or processing of coffee in Kenya, or cotton in many West African countries. Historically, extension has seldom been the motivation for formation of groups; but recent experience has shown that once extension starts working with groups, which have realized the value of extension advice, extension acts like a glue which binds the group together. There are many recent instances of groups having been formed around extension messages. Women's groups have almost always developed to adopt extension recommendations. In Somalia, groundnut was introduced as a new crop in Baidoa region. Farmers who started growing groundnut for the first time after receiving extension advice, tended to form a group, and exchange information among themselves.

5.104 Given the fact that groups achieve greater farmer coverage, quicker spread of technology, provide vital feedback to extension regarding relevance of recommended technologies to field situations, any kind of group is welcome. In West Bengal the Youth Clubs, in Assam the F.M.C's, in Gujarat, the Gram Vikas Mandals of the A.K.R.S.P, in Rajasthan the Kisan Mandal, in many States the Milk Unions, the Oilseeds Grower Societies, the Mahila Mandals could all be activated for agricultural extension. In still other areas NGO developed village institutions could also be mobilized for extension.

5.105 However, where farmers organizations do not exist the importance of the Governments' facilitative role in fostering farmers associations cannot be overstressed. It is not always necessary that extension actively promotes farmers groups. Extension work with farmers involves persuading them to try new practices on small portions of their fields and inviting surrounding farmers to participate in discussion on the adoption of these practices. In due course, if extension works consistently with the same group of farmers who find the recommendations useful, a certain bond is established among the farmers and they evolve into a group.

5.106 In this regard the crucial contribution which NGO's can make should not be minimised. Given their organizational and social capacity building strengths, their support should be enlisted in mobilising farmers into organized groups for voluntary group action for self reliance and self development. They could be involved in training extension workers in communication methods, particularly participatory approaches.

Extension Volunteers

5.107 One conclusion which arises out of the analysis of conventional rainfed extension approaches is the extremely high cost of delivery as well as the very thin coverage on account of the increasingly wider ratios of Extension Workers to farm families. Against the World Bank norm of 1 VEW per 1000 farm families, the actual ratio in several States is around 1 : 2500 at present.

5.108 Given the high rate of growth of population and the need to therefore constantly increase the number of extension personnel to maintain a reasonable ratio of extension worker to farm families under the T&V system of extension, and the inability of the State Governments to fill up even existing vacant positions, what to talk of creating new positions, on account of severe resource constraints faced by them, a time has come for us to move away from conventional approaches of extension. It is in this context that the concept of Extension Volunteers drawn from the community needs to be operationalized. This concept is not new in the sense that it is already being used as a delivery mechanism in the NGO sector in a big way – the A.K.R.S.P. in Gujarat has been one of the pioneers in this regard. Even in State systems examples can be found. For instance, Gopals in Rajasthan are Community Livestock Volunteers, who are trained by the Animal Husbandry Department in A.I. and Veterinary care, provided a Nitrogen flask and frozen semen, and then encouraged to work in their community with the freedom to charge for the cost of services rendered by them. They are also paid an honorarium on a tapering basis for three years by the end of which they are expected to become self-sufficient.

5.109 This approach, apart from being cost effective, has another advantage in that it promotes the involvement and participation of the local community in agricultural development. Farmers are more likely to be stimulated by one of their own members in identifying their needs for improvement, in taking the necessary initiatives, and in demanding from the Government the technical and economic assistance required to sustain their efforts. In order to achieve this aim, the local community or organization would be asked to identify activists from amongst themselves. These volunteers would be provided periodic training by the extension agency and may or may not be paid honorarium. In case such volunteers are not forthcoming without financial incentives then only a token honorarium on a tapering basis and that too routed through the local farmers' organization may be made admissible.

5.110 The main advantage of this approach would be to create in the village a receptivity to change and a sense of participation. This would also enable extension effort to be more focused and relevant to the needs of the farming community.

Gender Differentiated Extension

5.111 Women as a target group are generally neglected by the extension services. Though this neglect is largely the result of deeply rooted attitudes to the role of women in traditional societies, these prejudices should not be allowed to perpetuate a situation that is an obstacle to progress. In several parts of the country men have little interest in agricultural work and women are the real farmers. Most of the agricultural operations are performed by women. In the circumstances, not especially addressing their concerns or addressing them through male farmers is not a justifiable strategy.

5.112 Therefore extension should re-orient itself to directly addressing the needs of the women farmers. Whereas in several parts of South India, the public extension system has a fairly large number of women extension workers, this sensitivity is more likely to be present (though this is not to suggest that this is invariably true). Where, however, women extension workers are conspicuous by their absence, efforts to encourage their entry into the public extension system should be made. In all situations, however, lady extension volunteers from the community should be identified to serve the role of social animators.

Involvement of Rural Youth

5.113 Similarly, the potentialities of youth movements and organizations as an educational factor can be exploited for extension purpose. Experience in West Bengal has shown that rural youth is very receptive to new ideas and can make a considerable contribution in influencing the attitudes of the older generation. Unemployed rural youth with some level of school education can be given vocational training of one or two years duration in agriculture and allied activities with the purpose of establishing themselves in income generating agricultural activities in the rural areas. They would also serve as extension volunteers. The strategy of utilising youth clubs or associations, wherever existing, has already been alluded to in the earlier recommendation on group approaches.

II.3 High Value Commercial Agriculture

5.114 High value commercial agriculture is specialized, diversified and localized. With the advent of export-led growth strategies and setting up of agricultural export zones, many organized private sector players have signed collaborative agreements with multi-national corporations (MNC) – technical and/or equity participation – with buy back arrangements floriculture, aquaculture, orchards, high value exotic vegetables, agro-processing are some of the newly emerging export enterprises. At the same time several Indian business houses have signed memoranda of understanding with biotechnology groups for hybrid seeds, tissue culture, etc.

5.115 One important feature of the export-led high value commercial agriculture is that the foreign collaborators bring technology to the host concern. The farmers operate as contract growers. The private sector agent ensures that the domestic and export commitments are met at a cost acceptable to the company and farmers. The extension concern in this case, is to produce a product of a quality acceptable to the export trade. In most cases, the company undertakes to provide all necessary resources, both financial and otherwise, so that the crop plans are fully implemented without unnecessary delays. All objectives and targets are quantified; thus every extension agent is clearly accountable to his superiors, and the reasons and susbsequent rewards or penalties are known to everyone. Farmers receive prompt payment for their produce.

5.116 Public sector extension and research can never match the efficiency and effectiveness of such profit-centre/responsibility-centre oriented extension and research.

5.117 In export-led or value addition processing led-crops, the research concern is with product designs that are of uniform quality. For instance, ITC Agro Business Division, set up a research based hybrid seed complex to produce sunflower seeds with uniformly high oil content 42 per cent and more. The concern here was to ensure high capacity utilization and greater oil recovery for its Sunflower mill. Similarly, Pepsi International brought its tomato and potato varieties for distribution among contract growers, with arrangements for buy back of produce which meets its processing requirement. Although, there are a large number of Indian bred tomato and potato varieties, none could stand the processing requirements of Pepsi. The private sector is playing an important research and extension role in such areas, as it is easy to apportion profit.

5.118 Growers' associations, such as Grape Growers' Association in Maharashtra or United Planters' Association of South India at Ooty are undertaking research and extension responsibilities for their members.

5.119 The public extension and research system has a very limited role in the high value commercial agriculture. Even peasant-oriented, intensive but market responsive and high value vegetable and floriculture cultivation depend on research based seeds and advice of private sector. The public extension may play a limited promotional role for this segment.

5.120 In the next chapter outline for pilot projects have been developed for small scale field testing of the preferred options for irrigated and rainfed situations in a few selected blocks. These pilot projects would have to be carefully monitored for effecting mid-course corrections as and when necessary.

NOTES

1. In Irrigated Tracts of India, SAUs have placed a team of 4-7 scientists at the district level with transfer of technology responsibilities.
2. It is proposed to strengthen a block level extension team through redeployment to undertake adaptive research in collaboration with researcher and farmer. Different options are discussed under the extension.

6

PILOT PROJECTS ON FARMER PARTICIPATORY RESEARCH AND EXTENSION SYSTEMS IN IRRIGATED AREAS AND RAINFED FARMING SITUATIONS

IRRIGATED AREAS

1.0 Objective

To reorient and make the existing agriculture research and extension systems more farmer participatory.

2.0 Strategy

2.01 *Extension*

- By making extension more responsive to farmers' needs.
- By promoting group approaches.
- By meeting specialised needs of farmers and farmers groups.
- By making system farmer driven and accountable.

2.02 *Research*

- By involving farmers actively in the process of generation of need based technology i.e. in problem diagnosis, prioritization, planning, designing, implementation, evaluation and feed-back.
- By constituting a management committee at Zonal Level comprising representatives from Extension, Farmers Groups,

NGOs, to determine the research agenda and supervise and monitor its implementation. The committee would also be charged with the responsibility of planning, guiding and monitoring extension.

— By introducing the concept of contractual research.

— By organising on farm research and adaptive trials on farmer's fields with funding by and active involvement of extension.

— By processing technological information based on research results for use of extension.

3.0 Project Area

The project would operate in one Block of a district in a State for a period of 2-3 years with only spill over liabilities of activities started in the second year to be provided in the third year.

4.0 Project Description

Participatory Agriculture Extension Plans

4.01 A plan to promote farmer participatory agriculture development would be formulated by block extension team in collaboration with farmers organization/groups. The plan would focus on sustainability issues and indicators of non-sustainability. It would identify major second generation problems, their magnitude arising as a result of adoption of technology and broad strategy to solve them through group action. The plan would also contain the future extension thrust areas.

These plans would be prepared for homogeneous groups of farmers indentified as per their needs/special problems.

4.02 The list of problems, constraints, issues would be screened by district scientists to identify technological and extension gaps. The technological gaps would be passed on to researchers for formulation of research projects in consultation with farmers organization or group. Similarly, the team would develop an action plan in consultation with district scientists to bridge extension gaps.

4.03 Joint Research and Extension Management Committee at the ZRS level would prioritize the problems. The guiding criteria would be (a) problem affecting largest number/area or intensity of losses occurring would gain precedence and (b) availability of funds at ZRS.

4.04 The designing of the experimentation on research station, farmers fields and adaptive trials on farmers field would be done by researchers. In case of the latter two, farmers and extensionists would actively participate.

4.05 Implementation of on-station research would be carried out as per design and plan developed by researchers. On farm research and adaptive trials would be conducted jointly by scientist team, extensionists block level team and farmers.

4.06 Monitoring and Evaluation would be done jointly by district scientists, block extensionists and farmers at different stages.

4.07 Feedback would be provided to researchers based on joint evaluation for information processing for dissemination or further research.

Extension Issues

4.08 *Team Approach*

Extension team at block level and scientists team at district level would interact with farmers organization/groups.

4.09 They would identify existing farmers organisations for technology dissemination. If no farmers organisation exists, team would facilitate creation of farmers organisations/special interest groups (having common problems). The assistance of credible NGOs operating in the block would be sought for this purpose.

4.10 The team would interact with farmers organisations/groups as per schedule prepared by them. Appropriate extension aids would be used during interaction.

4.11 The block team in collaboration with district scientists would organise clinics at Agriculture Information Centre. They would also be available at the centre for consultation, advice to farmers on days and time as fixed by them and widely published.

4.12 *Planning — Bottom Up*

Participatory agricultural extension would require bottom-up planning on the basis of needs identified. The extensionists would work out demand for types of inputs, minikits, demonstrations etc. rather than imposed with targets from the State, district and divisional levels. Instead of acting as a "delivery system", the public extension system, henceforth, can help to organize local people to develop their own "acquisition" systems, to acquire from the external environment those technologies and services that they want. The approach would be to mobilise farmers for group action for area based problems. The management of inputs, subsidies etc. should be on area basis and it should be a one time operation covering the entire group.

4.13 *Flexible Funding of Innovative Farmer Participatory Approaches*

Provision of untied funds be made in the district level allocation of funds to undertake innovative projects based on group needs.

4.14 In every centrally sponsored programme, it is desirable to have a component of promoting farmer participatory research and extension programmes.

A certain percentage of funds may be earmarked for expenditure on such items as:

(a) On farm research, adaptive trials on farmers fields.
(b) Technological demonstrations to promote group action.
(c) Technology dissemination through mass media.
(d) Institution building
(e) Human resource development.

4.15 A new programme could also be developed at national level to encourage State Governments to promote farmer participatory approach in research and extension systems.

4.16 *Contractual Research*

The concerned government departments need to provide lumpsum funds to ZRS for undertaking need based research as diagnosed and prioritized through the process of farmers participation described in paragraphs 4.01 to 4.08. Funds would be made available through contractual agreement between concerned government department and ZRS. Similarly, contractual research could also be conducted by entering into agreement with private and corporate sectors.

5.0 Policy Intervention

The Department of Agriculture of the State Government would require the following policy decisions:

(a) Constitution of a block level team by deploying VEWs and SMSs.
(b) Team approach in dissemination of technology as per needs of farmers.
(c) The farmers groups will be the unit of interaction.
(d) Active involvement of University Scientists in Extension.
(e) The present methodology of T&V would be modified as per methodology outlined in the Project.

6.0 Financial Implications

Assumptions

6.01 The financial implications are being worked out on incremental basis with a block as unit for a period of 3 years.

The block level team comprising 1 SMS, 2 AEOs and 10 VEWs would be responsible for extension work with farmers/farmers groups in one block comprising of 100 villages each having a population of 100 farmers on an average i.e., 10,000 farmers.

The project would be implemented in one block each in 4 different States.

6.02 It is assumed that 25 farmers organizations would be activated in the first year and another 50 in the second year.

6.03 No new staff would be employed but extension and research would be carried out by deployment of existing staff and alternate role assignment. Similarly, mobility would be ensured by providing existing

Physical Programme for One Block

Item/Activity	Year I	Year II	Year III
(a) Number of farmers organizations/ groups to be activated.	25	50	Nil
(b) Number of training programmes for block level team	1	1	
(c) Number of sensitivity training for supervisory level extension functionaries or scientists.	1		

Financial Programme for One Block

(Rupees in Lakhs)

Item/Activity	Year I	Year II	Year III
(a) Lumpsum amount for servicing farmers groups/organizations (POL, media support, meeting expenses).	1.25	3.75	Nil
(b) Cost of Training			
(i) Block team Rs. 100 per trainee for institutional training for 15 days	0.20	0.30	Nil
(ii) Sensitivity training for supervisory extensionists and scientists at the rate of Rs. 200 per day for 7 days.	0.25	Nil	Nil
(c) Lumpsum propulsion charges to ZRS for visits of District scientists and researchers.	1.00	1.20	1.30
(d) Propulsion charge for block extensionists.	0.10	0.15	0.20
(e) Contractual Research	2.00	3.00	
(f) Flexible Fund	2.00	3.00	
(g) Unforeseen contingent expenses	2.00	2.00	
(h) Setting up Agriculture Information Centre and conduct of clinics.	0.50	0.10	0.10
Total	9.30	13.20	1.60
For One Block	= Rs. 24.40 lakhs		
For Four Blocks	= Rs. 97.60 lakhs.		

vehicles or paying hire charges.

RAINFED FARMING SITUATIONS

1.0 Objective

To reorient the existing agricultural research and extension system to make it more farmer participatory.

2.0 Strategy

2.01 *Extension*

— By making extension more need based by adopting a farming systems approach.
— By encouraging group based approaches.
— By involving community volunteers in extension.
— By targeting women and youth in extension effort.

2.02 *Research*

— By involving farmers actively in the process of generation of need based technology, i.e., in problem diagnosis, problem prioritization, planning, designing, implementation, evaluation and feedback.
— By constituting a Joint Management Committee at the Z. R. S. level comprising representatives from Extension, farmers' groups and NGOs which would be charged with the responsibility of determining the Research Agenda, supervise and monitor its implementation. The Committee would also be responsible for planning, guiding and monitoring extension.
— By introducing the concept of contractual research.
— By organizing on-farm research and Adaptive Trials on farmers fields with funding by and active involvement of extension.

3.0 Project Area and Duration

The project would operate in one block of a district in a State for a period of two to three years with only spill over liabilities of activities started in the second year to be provided in third year. While selecting the Block, preference may be accorded for an operational area of an existing and credible NGO.

4.0 Project Description

Participatory Agricultural Development Plans

4.01 A participatory agricultural development plan would be forumulated by the VEW in collaboration with farmers organizations at the village level. The plan would be a brief document, analysing the resource position of the village, its cropping pattern, rainfall trends etc. It would identify the most important constraints affecting increased agricultural production, the major potentialities and, arising from the aforesaid information, the broad thrust or plan for future agricultural development.

4.02 These plans would be prepared for each village and then homogeneous problem areas delineated.

4.03 Thereafter, the exhaustive list of problems, constraints, issues generated would be screened and technology and extension problems separated. The former would be passed on to the Z. R. S. for formulation of research projects in consultation with farmers' organisation in one or more representative villages in the homogeneous situations delineated. Similarly, extension problems, constraints, issues would be passed on to the relevant line departments for development of extension packages for dissemination.

Technology Issues

4.04 After problem formulation, problem prioritization would be done in a Joint Research and Extension Management Committee at the Z.R.S. level comprising extensionists, researchers, farmers and NGOs. The criteria to be adopted for prioritization would be twofold:

(a) Problems affecting largest number of farmers or inflicting maximum losses would gain precedence.
(b) Flexible funds available with the Z. R. S. would be kept in mind.

4.05 Then the planning and design of the projects selected would be undertaken, again jointly by extensionists, researchers and farmer organizations in one or more of the representative villages. Possibilities of experimentation on farmers' fields would be explored and appropriate sites selected.

4.06 Implementation would be carried out as per design and jointly supervised by extensionists, researchers and farmers. Depending on nature of prioritized problems, studies can be initiated either at the experiment station or on farmers fields. Where it is on the former, the recommendations will await findings of on-farm research and adaptive trials.

4.07 Feedback would be obtained periodically from farmers and extensionists.

4.08 Monitoring of the programme implementation vis-a-vis design would be performed by a joint team of extensionists and researchers.

4.09 At different stages of on-farm research, adaptive trials etc., farmers would be provided an opportunity to evaluate the experimentation employing their own criteria. Evaluation of the project vis-a-vis its objectives would also be made by a team comprising extensionists and researchers.

4.10 Based on the evaluation and the feedback generated, either the research agenda would be reformulated or extension messages formulated for dissemination.

Extension Issues

4.11 *Broadbased Extension*

Before dissemination of extension messages, a policy decision would be taken at the level of the State Government to involve the relevant line departments in the technology diffusion process. The process suggested in paras 4.01 to 4.03 could be employed for identification of the relevant departments in different parts of the State, if considered useful. Upon identification of the relevant line departments, the modalities of broad-basing would be worked out. Modalities would include the linkage mechanisms, the training methodology etc.

Farmers Organizations

4.12. User groups, including farmers organization, farmer clubs, commodity growers association, young farmer clubs, women farmer groups, special interest groups and the like are effective institutional devices in the creation of client driven agricultural research and extension system. Extension can play an important role as a catalyst in facilitating organisation of these farmer groups. These groups can then become important mechanisms in articulating specific research and extension needs, in facilitating horizontal diffusion, in accelerating technology dissemination and in developing technical and managerial skills and knowledge of farmers who participate in these organisations.

The broadbased extension worker would be required to identify existing farmers' organisations at the village level for technology dissemination and feedback. In case no organization exists, the VEW, with assistance of AEO, would facilitate creation of village organizations to receive extension packages and provide feedback. The service of credible NGO's operating in the area could also be drafted.

4.13 Henceforth VEW would only interact with the farmers' organization at a prefixed venue on a prefixed date and time. Wherever possible, AEO would also accompany.

4.14 The schedule of trainings and visits would have to be reworked keeping in mind the availability of extension personnel, availability of SMS from other departments, the number of farmers' organizations and the qualitative content of extension packages.

4.15 It would be made compulsory for VEW, while interacting with farmers' organizations, to use atleast one of the following aids in every interaction session with farmers' organizations—Overhead Projector, Slide Projector, VCR, Flip Charts, other Charts etc., by obtaining the same from district or sub-district office or on hire

Farm Extension Volunteers

4.16 Another cost effective method of promoting farmers' involvement in extension is the concept of Extension Volunteers. Potential innovators and entrepreneurs possessing land and preferably not belonging to the rural elite would be identified by the farmers' organizations as extension volunteers. There could be more than one volunteer from a village of which at least one should belong to women or youth groups. These volunteers would be provided regular training on a periodic basis—both institutional and peripatetic—in appropriate technical skills, technology and communication methods, and provided a token monthly honorarium on a tapering basis for 2 to 3 years. Additional help in mobilising loans for setting up service facility for the volunteers i.e., input distribution, custom hiring etc. is envisaged. This payment would be routed through the farmers' organization in order to increase accountability of the volunteer to the farmers' organization. This Volunteer would also serve as a link between conventional extension structures and farmers' organization. This participatory strategy can overcome the problem of vacant positions also by having larger numbers of volunteers whose background is appropriate to local needs.

4.17 Special Training Pogrammes would be evolved for women and youth separately and batches of trainees to be identified by the farmers organizations would be trained in modern technology and communication methods relevant to their needs.

4.18 *Bottom-up Planning*

Instead of targets being handed down arbitrarily from the State down, the participatory agricultural development plan would help reverse the process of planning. On the basis of need identification made by the farmers' organizations, the VEW could place demands on the system to deliver inputs, mini-kits, demonstrations etc. under various programmes at his doorstep. For instance if the farmers have indicated an interest in a demonstration for alternative crop for Jawar in Kharif, or in a Sprinkler demonstration set for generating awareness of the advantages of improved water management practices, or they have shown their willingness to adopt Plant Protection measures for which one time supply of sprayers or dusters and subsidised P. P Chemicals is required, then these requirements would be met from different programmes. Instead of acting as a delivery system, the public extension system, henceforth, can help to organise local people to develop their own "acquisition" systems, to acquire from the external environment those technologies and services that they want. Through this process, "control" of extension through programme formulation and extension volunteers is passed on to the ultimate clientele. In other words, the implementation of programmes should be 'demand' driven rather than 'supply' driven.

Flexible Funding of Innovative Approaches/Farmer Participation

4.19 Also, concept of flexible funding would need to be introduced. In every State department budget there should be an amount roughly equal to 5 per cent of the State Plan budget which should remain untied to any programme. This amount should be available for taking up small local pilot programmes on the basis of need expressed by local communities and conveyed through VEW. This amount, which should be placed at the disposal of the Secretary of the Administrative Department would not, however, be available for staff, buildings, purchase of vehicles or capital expenditure. It would only be for funding small innovative ideas for a period not exceeding one year, provided the idea or innovation cannot be taken up under an ongoing programme.

4.20 In every central sector and centrally sponsored programme it is desirable to have a component of promoting greater farmer participation in agricultural research and extension systems. The amount to be provided need not come by way of any increase in budgetary outlay but by merely earmarking 10 per cent of the current outlays for funding farmer participatory programmes or approaches. An illustrative list of the types of items of expenditure, provision for which could be made in the afore-mentioned component are as follows:

(1) Technology dissemination through mass media – Organising Fairs for farmers, field days, printing of pamphlets, posters, literature, charts, etc.
(2) Horizontal Diffusion – organising farmer to farmer visits within and even outside the State.
(3) On farm research, Adaptive Trials on farmers' fields
(4) Institution building
(5) Human Resource Development

4.21 Infact the Government of India could also introduce a new National Programme on encouraging greater farmer participation in Agricultural Research and Extension systems. This programme can create an enabling environment for States to take initiatives in reorienting their research and extension services to make them more farmer accountable and farmer responsive. A lot of flexibility should be provided to encourage experimentation of new ideas and approaches and field testing them.

4.22 *Contractual Research*

Within the budget of the agriculture and other line departments, State Seeds Corporation, Agro Industries Corporation, etc. a lumpsum provision for contractual research should be kept. This fund would be utilised for research problems thrown up by the participatory process described in paragraphs 3.01 to 3.09. Funds would be released on the basis of a contractual agreement between the agriculture department/ concerned line department and the Agriculture University (ZRS). Monitoring and evaluation would be done jointly by Extensionists, Researchers and Farmers through the Joint Management Committee of the ZRS and further releases made upon satisfactory progress achieved against pre-set performance indicators.

5.0 Accountability and Responsiveness

5.01 The participatory approach coupled with farmers organization and extension volunteers ensure that programme planning is controlled locally. VEWs agenda is formulated locally. Participation by representatives of extension, research and NGO's will contribute to its success and legitimacy. But the decentralization of programme planning control allows priorities to vary as per the local needs.

5.02 The VEW is provided backstop technical and financial support to implement locally decided programmes. Responsibility is matched with decentralized decision making authority. Extension volunteers supplement VEW's, the extension structure assists the farm extension volunteers to implement locally designed extension programmes.

5.03 In addition to the hierarchical authority, farmers organization is involved in monitoring and evaluation of extension staff. Thus the current monitoring indicators, such as "number of visits made to farms", the "number of meetings arranged and the "number of demonstrations organized" may change to "impact per person", impact per unit area", or "effectiveness in solving some of the prioritized problems."

6.0 Incentives and Rewards

6.01 Appropriate performance linked incentives and rewards systems for VEW's in which local farmers organization play a role is to be introduced through a graded system of rewards. The graded system should reinforce VEW's farmer participatory behaviour and the impact thereof. Under this system, on the basis of unanimous resolutions of farmers' organizations, some VEW's, would qualify for citations of merit, a few, say 10 per cent, would be eligible for a monetary incentive of Rs. 500; and the best VEW in the block, an award along with cash prize of Rs. 1,000.

7.0 Policy Intervention

The State Government would be required to:

(a) take a decision as to which 2 or 3 major land based needs of farmers can be brought within the ambit of broadbasing and the pace, extent and modalities of broadbasing.

(b) constitute Committees for coordination and overseeing the progress of broadbasing.

(c) take a decision on making the dominant enterprise department of an area as the lead department and place all field extension personnel of the Agricultural Department under the administrative control of the lead department.

(d) accept the concept of Extension Volunteers both as a cost-effective measure as well as to increase involvement of the local community.

(e) approve the concepts of flexible funding, contract research and awards and incentives linked to performance.

8.0 Financial Implications

Assumptions

8.01 The financial implications are being worked out on an incremental basis with the district as a unit and for a period of two-three years. It has been assumed that there are about 1000 villages in a district organized into 10 Panchayat Samitis or Development Blocks. In other words, each Block would have approximately 100 villages within its jurisdiction. The pilot project is to be implemented in one block of one district each in atleast four different States.

8.02 It has also been assumed that the ratio of VEW to villages is 1:6 and AEO to VEW is also 1:6. On the basis of this assumption, the availability of VEW's and AEO's in a Block has been assumed as 16 and 3 respectively.

8.03 It has further been assumed that only 25 farmers organizations would be activated in the first year and another 50 in the second year.

8.04 An additional assumption is that in the first year 50 Village Extension Volunteers would be identified and in the second year another 100 volunteers would come forward.

8.05 It is also being assumed that there would not be any need for incremental staff, vehicles etc. as this is a pilot project of only 2 years duration and is to be implemented only by re-deployment, alternative role assignment etc.

8.06 *Physical Programme—For One Block*

Item/Activity	Year I	Year II	Year III
(a) Number of farmers' organizations to be activated and serviced	25	50	Nil

(Contd.)

Item/Activity	Year I	Year II	Year III
(b) Number of Extension Volunteers to be identified and trained.	50	100+50=150	Nil
(c) Orientation programme for State level Nodal Officer, SMS of various line deptts. and the district/sub-district level officers of all line departments, NGO's operational in the area. (3 days duration)	1	1	
(d) Number of Training Programmes			
(i) For SMS's of all deptts. (3 days) Batch of 10	2	2	—
(ii) For AEO's/VEW's in Broad-based extension and Group Communication methods (Batch of 20)			
— Instnl. (15 days)	1	1	
— Field (1 Day)	10	10	
(iii) For Village Extension Volunteers (Batch of 25)			
— Instnl. (7 days)	2	4	
Field (2 days)	12	24+12=36	

8.07 *Financial Programme—for One Block*

(Rs. in lacs)

Item of Expenditure	Year I	Year II	Year III
(a) Lumpsum amount for servicing Farmers Organisations (P.O.L., Tea, Publicity, Media support, etc.) @ Rs. 5000 per Organization per year.	1.25	2.50+1.25	
(b) Payment of honorarium per annum to Village Extension Volunteers @ Rs. 600 in 1st year, Rs. 400 in 2nd year.	3.60	9.60	4.80
(c) Orientation programme for State level Nodal Officer, SMS' of various line deptts. and the district/sub-district level officers of all line departments, NGO's operational in the area.	0.10	0.10	
(d) Cost of training programmes			
(i) Institutional Training For SMS's of all line deptts. at Rs. 200 per day	0.12	0.12	

(Contd.)

Item/Activity	Year I	Year II	Year III
(ii) For AEO's/VEW's @ Rs. 100 per trainee per day for Instnl. training	0.30	0.30	
and @ Rs. 30 per trainee per day for field training	0.06	0.12	0.12
(iii) For Village Extension Volunteers @ Rs. 100 per trainee per day for instnl. trg. and	0.70	1.40	
@ Rs. 30 per trainee per day for field trg.	0.18	0.36+0.18	0.54
(e) Lumpsum propulsion charges to Z.R.S. for field visits in connection with Farmer Participatory Work	0.20	0.25	0.30
(f) Lumpsum Propulsion charges for Block level officials of the Agril. Department.	0.10	0.15	0.20
(g) Flexible fund provision	2.00	3.00	
(h) Contract Research Provisions	2.00	3.00	
(i) Awards & incentives	0.02	0.02	
(j) Amount for contingent and unforeseen expenses	2.00	3.00	
Grand Total	12.63	25.35	5.96

For one Block = Rs. 43.94 lacs or Rs. 44 lacs.

For four Blocks = Rs. 176 lacs.

8.08 The amounts shown in para 8.07 are only indicative. Each State would have the flexibility of making suitable changes, wherever necessary, on the basis of their assessments. Diversion from one component to another and provisioning for new items from out of contingency would be allowed, subject to sanction of a State level Sanctioning Committee to be constituted under the chairmanship of the Secretary, Agriculture and including Heads of executing line departments, Director, Research and Extension of the SAU, the Associate Director, ZRS, the Joint and Deputy Director having jurisdiction, representatives of the NGO which is operational in the area and representatives of farmers' organisations. The size of the Committee should not exceed 15.

ANNEXURES

ANNEXURE I

PROFILE OF CONSULTANTS

K. C. John is Programme Consultant, Sustainable Agriculture Programme at the New Delhi Office of the Ford Foundation. He received his doctorate in management from Indian Institute of Management, Ahmedabad. He has held teaching, research positions at the Indian Council of Agricultural Research and the Indian Institute of Management, Ahmedabad. Dr. John has designed and facilitated implementation of farmer participatory techno-logy development and dissemination programmes at several State Agricultural Universities and Non-Governmental Organizations in India. Besides, he has overseen methodology development, information exchange and policy dialogue activities to institutionalize participatory approaches in the National Agricultural Research Systems in South Asia. His principal intellectual and practical concern is in the fields of institutional analysis and design, strategic planning, livelihood strategies and sustainable resource management. Dr. John has published extensively on participatory development, institutional change and management of agricultural research/extension.

Devendra Kumar Sharma obtained M. S. (Rural Sociology) and Ph.D. (Extension Education) from Cornell University, U.S.A. He visited U.S.A. under TCM Programme in 1956. In 1964 he was awarded Ford Foundation Fellowship for advanced training in Extension. In 1986 he visited U.K., France, Holland, Denmark, Switzerland, Italy, Egypt, Japan to study agriculture extension systems under the auspices of International Development Programme of Cornell University. He started his service career as Lecturer, College of Agriculture Nagpur in 1952. He worked as Professor, Head of the Department, Dean, Director Extension Education at JNKVV, Jabalpur and Zonal Coordinator, ICAR Lab to Land Programme, Western Zone. Dr. Sharma worked as Advisor, Agricultural Extension to Ministry of Agriculture, Government of Ethiopia under World Bank consultancy. He was founder Vice

Chancellor, IGKVV, Raipur and also worked as Vice Chancellor, JNKVV, Jabalpur. He served on various academic bodies of several Universities, ICAR Scientific panels, State and Central Government Committees. Dr. Sharma received "Excellence in Extension Award" in year 1991 for outstanding contribution in the field of Agriculture Extension.

C. S. Rajan is an I.A.S. Officer of the 1978 batch, who has been serving the Government of Rajasthan as Director of Agriculture (since May 1989) as well as Managing Director, Rajasthan State Seeds Corporation (since April 1990). His stewardship was marked by a phenomenal growth in agricultural production, Rajasthan emerging as the leading State in oil-seed production, fertilizer consumption growth and innovations in Agriculture. He presented a paper on Environmental Impact Assessment in Agricultural Projects in India in an International Seminar in Tokyo in 1990. He attended a training programme on Organizational and Management Development for senior Administrators in Agriculture and Rural Development in N.T.L., Washington, U.SA. in May-June, 1992. Shri Rajan also represented the Government of Rajasthan in successfully negotiating a Rs. 400 crore Agricultural Development Project with the World Bank in Washington in October 1992.

Charanjit Singh served as Director of Agriculture, Himachal Pradesh from May 1987 to February, 1991. He obtained his Doctorate from Technical University, Berlin and was a recipient of Scholarship from German Academic Exchange Service. He was consultant to the U.S.A.I.D. sponsored Hill Areas Land and Water Development Project, Himachal Pradesh. Singh had also occupied positions like Joint Director, Agriculture and Project Officer, Indo-German Agriculture Project. He was instrumental in introducing farmer managed small irrigation schemes and organising Krishak Vikas Sanghs in Himachal Pradesh. In addition, he also prepared State agriculture extension and water management plans for irrigated areas of Himachal Pradesh. He presented a paper in European Society of Agronomy Conference at Berlin. Presently he is a member, Board of Management and Extension Council, Himachal Pradesh Krishi Vishwa-vidyalaya, Palampur. Dr. Singh has visited Sri Lanka, Philippines, Taiwan, Germany, Singapore, Italy, Holland, England and Canada to study the agricultural extension and allied agricultural development works.

S. K. Arora, Director, National Institute of Agricultural Extension

Management (since July 1992), is an IAS Officer of the 1971 batch from Andhra Pradesh Cadre. He read Political Science at the masters level in Punjab University and completed MA in Rural Development from University of East Anglia, Norwich, U.K. He began his career as Lecturer, Political Science at Chandigarh before joining the Indian Administrative Service. Shri Arora combines in himself a unique blend of experiences – implementation of agriculture and rural development projects at the district level; Policy making and overseeing responsibilities as Secretary in State Departments of Agriculture, Panchayati Raj and Rural Development; and as Commissioner and Registrar, Cooperative Societies; business management as Chairman and Managing Director, A. P., MARKFED; Chairman, A. P. State Seeds Corporation and General Manager, National Cooperative Development Corporation. In the field of post-experience training, he was instrumental in setting up 11 State-level Cooperative Staff Training Institutes and developing curricula and course materials.

ANNEXURE II

STUDY ON INTRODUCING FARMER PARTICIPATION INTO PLANNING AND MANAGING EXTENSION AND RESEARCH SYSTEMS IN INDIA

TERMS OF REFERENCE

Background

Under a series of State and national agricultural extension projects, starting in 1977, 17 State extension services have been established by State Governments with support from the World Bank. A physical infrastructure of staff housing, training facilities, and vehicles is established and about 88,000 fulltime extension staff are positioned in the field. The work of this considerable staff is well organized and on the whole well disciplined; every month in India about 400 joint extension research workshops and 800 field staff training sessions are held. Village Extension Workers make regular visits to farmers' fields according to a fixed schedule. linkages between extension staff and the research systems of the 26 State Agricultural Universities are much better than before and as a result the views of extension workers are taken into account by agricultural scientists in planning and research.

Under two National Agricultural Research Projects, which started in 1980, the linkages between development departments, especially agriculture, have also been greatly strengthened. Under NARP, agricultural research has been substantially decentralized and made more location-specific. 120 agro-climatic research stations (with substations) have been developed, each staffed with a multidisciplinary group of scientists. 240 zonal research advisory meetings are held annually. These are attended by SAU representatives, extension staff, and representatives of line departments. These committees evaluate and plan on station and on farm zonal research programmes. Scientists of zonal stations also participate at extension monthly workshops as trainers, conduct diagnostic field visits with extension staff, organise station field days for farmers, and work with extension on farm trial programmes.

Over the past 10-15 years a strong and sustained working relationship has been built up between agricultural extension and research services in most States. Although the quality of both extension and research operations varies considerably between and in States there is a general appreciation by GOI and State governments of the institutional developments achieved under the extension and research projects. There is a strong foundation in the field to work with and build upon.

The Issues

It has been generally accepted both in the cases of extension and research that there is considerable scope to improve quality, management and cost effectiveness. Some measures to address these have already beeen identified and are being implemented to varying degrees. A major outstanding challenge both for extension and research is how, in the Indian scenario, public sector extension and research services can be made more client (farmer) driven and controlled?. How can real (non-cosmetic) farmer participation be introduced into planning and management of extension and research in a manner that is replicable on a State/nation-wide basis. How can extension and research staff be made more accountable to the client farmer when employed reported on and paid by a public organisation such as a State development department or a State agricultural university. What changes, incentives, or management tools could be used to encourage such a trend. How can farmer or community opinion/evaluation of a public servant's performance be safely and fairly obtained and recorded and used as a factor in determining a public employee's rewards. These are but a few examples of the questions that need to be posed and answered.

With the exception of a few limited and a typical special cash crop situations (grapes, tea etc.) this has not been achieved on any scale in India. Little is being done in the field to test options and generally there is a dearth of practical ideas on how to go about addressing this issue.

Study Objectives

The study's first objective is to develop one or a number of options on how stronger farmer participation can be introduced into extension and research management in India and how research scientists and extension workers can be made more accountable to the client community.

The existing structures cannot readily be dismantled so the options must be built round what is in place and take into consideration political and bureaucratic realities. The second objective is to recommend how and where the proposed options for introducing farmer participation can best be tested, on what scale, and how they may be rigorously and if possible empirically evaluated. Apart from presenting an analysis of the present situation and proposing options for improving farmer participation both in research and extension the consultants should also discuss and agree with GOI an implementation programme for testing the agreed and most promising options.

Methodology

The study team would hold discussions with a range of Government and non-Government persons involved or interested in the subjects, both at State and Central levels. The study would require intensive field work to review government and non-government experiences or farmer participation in extension research and current State research and extension operations in a sample of NARP agro-climatic zones/ extension districts would be selected, probably in three States representatives of three main situations, one southern, one eastern and one northern hilly. Apart from liaising with State government and SAU officials considerable time should be spent listening to views of farming communities. As farmer perceptions are expected to be gender and class sensitive a survey must cut across all social groups. A specially designed survey of farmer opinion and perception will be required to ground study analysis and findings. The purpose of this field work would be to find answers to the issues posed at para 4 above.

It is anticipated that the study would be conducted in three main parts:

(1) (a) Desk review of previous experience in farmer participation and control of public-sector extension and research and identification of those cases most likely to be relevant in the Indian situation. This work would be conducted by assessing already available library information and

experience; and

(b) Visit to, and analysis of the most promising case experiences of farmer participation and approaches identified above (1) (a) jointly by Consultants and MANAGE.

(2) Field work in India including the establishment and briefing of local consultant teams; discussion with research and extension officials and with projects or agencies already conducting participatory approaches in India; and conduct of survey of farmer opinion (mainly done by local consultants with supervision from ODI; three two-person local consultant teams for one month in each of the selected States would be constituted for this work).

(3) Preparation of draft report including analysis of findings and recommendations on participatory models worth testing in India; followed by discussions/workshops with GOI, and then a final report.

Time Schedule for Farmers' Participatory Study

January 28-31, 1993	Preliminary discussion at MANAGE
February 1-2, 1993	National Workshop
February 3-7, 1993	Finalise scope of study, locations and survey formats.
February 8-16, 1993	Field visits to ongoing projects.
February 17-21, 1993	Revise survey formats and field survey locations.
February 22-March 20, 1993	Field Survey.
March 20-28, 1993	Team-wise report preparation
March 29-April 4, 1993	Compilation of draft final report.
April 5-11, 1993	Discussion on draft final report and national workshop.
April 12-17, 1993	Final Report.

Terms and Conditions of Consultancy Assignment

The consultant will work as a member of a compact team between the dates of 28th January and 17th April, 1993. The majority of this time will be spent in field. However, adequate time shall be allowed for briefing, desk review, report preparation and debriefing in Hyderabad.

Specifically, the Consultant will have the following responsibilities:

> Critically review the existing experience in farmer participation and control of public sector extension and research in India, examine the strategic design choices of the most promising case experiences of farmer participation and approach through field visits and semi-structured interviewing, use this knowledge to formulate guidelines and design a field survey format in three selected agricultural situations in the country, participate in the field survey in one or two specific agricultural situations, examine the appropriateness of the existing extension programmes to the current and emerging needs of sustainable agricultural production in the selected agro-climatic zones, suggest institutional innovation, methodological improvements, modifications in current planning and management of extension and research linkages and orientation programmes in support of the client-driven and controlled public-sector extension services, develop multiple options for pilot scale experimentation in farmer-participatory extension and research approach within a structured period of time, define possible roles for external agencies in the process of change, provide costing, where possible, for support of the recommended pilot scale experimentation.

The Consultant will be responsible for compiling team reports and drawing up a Strategy Report. This will cover the points noted above. The draft report will be discussed with Government of India/State Governments research and extension services. Inputs from such discussions will be suitably incorporated in the Final Strategy Report.

ANNEXURE III

Extension Documentation No. 2

DESK REVIEW: NATIONAL INSTITUTE OF AGRICULTURAL EXTENSION MANAGEMENT, HYDERABAD

LINKING FARMERS, EXTENSION MANAGERS AND RESEARCHERS

Policy, Mechanism and Management

Introduction

Managers around the world have found that the linkages between farmers, extension workers and researchers are difficult to organise and sustain, particularly when addressing the needs of resource poor farmers. Past experience shows that weak links diminish the impact of agricultural research, impair performance and reduce efficiency.

The linkage problems are to be studied in the context of policy, institutional mechanisms and staff management issues. In practice, it is managers who make the difference between strong and weak links. Active management should provide leadership, maintain flexibility and responsiveness and ability to manage conflict.

This document is a selection of experiences discussed in the extension literature which are valuable reference points. It is hoped that in future such experiences will continue to be documented and the present document updated.

LINKAGE BETWEEN RESEARCH AND EXTENSION

Evidence of Benefits from an Integrated Rural Development Programme

— Small-scale mixed farming is the predominant source of agricultural production in Colombia's Nanno Highlands. The pre-1978 period was characterized by:

— Weak links between the main institutions involved in technology development and transfer;
— Poor and stagnant performance in the main agricultural activities of the region—dairy, wheat, guinea pigs, maize, beans and barley: the only important exception was potatoes.

In the mid-1970s, Colombia implemented the Integrated Rural Development Programme (DRI) to improve agricultural performance in the peasant sector. To achieve this goal, it was necessary to coordinate the activities of the core technology development and transfer institutions. This coordination focused mainly on wheat, dairy, and guinea pigs. Among the specific linkage mechanisms developed were:

— Regular meetings between the DRI and representatives from extension, training, input distribution agencies, and farmers, to assist the DRI in targeting, financing, and monitoring activities;
— monthly coordination meetings between the DRI and leaders of all institutions, to reinforce targeting and strengthen inter-institutional links;
— quarterly meétings, to identify technological problems and allocate resources accordingly;
— collaborative groups, to carry out specific research, and extension tasks at district level;
— the use of subject-matter specialists to coordinate adaptive research, training, extension programmes, and the production of extension materials;
— on-farm trials, conducted jointly by researchers, extension workers, and farmers;
— joint planning between researchers and extension workers, to increase the coverage and quality of extension activities.

The effects of the DRI began to be felt after 1978. In the 1980-85 period there were substantial increases in milk, wheat, and guinea pig production. Significantly, however, there was little increase in maize and bean yields, the two crops for which there had been far less integration between research and extension.

(Ref. Engel. 1990)

2. TECHNOLOGY DEVELOPMENT BY FARMERS

A Tradition of Planning Hundreds of Years

The need for farmer participation is often clouded by the metaphors used to describe the process by which technology is developed and delivered. Technology is often thought to be a single, uniform product that is generated by researchers and 'flows downstream' to farmers in a one-way, linear process. The fact that some technology remains 'on the shelf' is ascribed either to the failure of farmers to appreciate its benefits or to 'bottlenecks' in the transfer process. Farmers tend to be seen as passive recipients-users of technology developed by other people. At best, it is acknowledged that some feedback on farmers' reactions to a new technology is desirable in order to refine that technology, but this is likely to be regarded as a need for mere fine-tuning.

These metaphors are misleading, as well as condescending. Farmers are not passive consumers, but active problem solvers who in fact develop for themselves most of the technology they use. For many hundreds of years before today's national agricultural research systems were set up, farmers did their own research and, by integrating technology from different sources and continuing to adapt on their farms, they still do today. The technology used by farmers is a complex product undergoing constant change.

(Ref. Roling, 1989)

3. FARMERS' WORKSHOPS

An Opportunity for Researchers to Learn

The Innovative Farmers' Workshops that have become a central feature of the Extension and Research Project in Bangladesh arose out of a specific incident. In 1981, at the request of local extension workers, the leader of the project's research team visited an area in which farmers were reported to be harvesting intercropped potatoes twice from the same planning. The farmers had discovered how to open the furrows, harvest marketable tubers, and close the furrows again, without damaging the crops. The advantage of the practice was not that it led to a higher potato yield but that it raised incomes, since the early potatoes could be sold at a high price.

The visitors considered the innovation an excellent one. In 1982 a two-day Innovative Farmers' Workshop – the first of its kind in Bangladesh—was organised to present the idea to a wider audience. A second workshop was held the following year, featuring different innovations. These workshops are now common and a total of 43 farmer innovations have been presented.

Each workshop lasts for several days, but does not cover more than two innovations. The number of participants is usually restricted to 30, to facilitate good discussion. The atmosphere is kept informal, so that farmers do not feel shy or nervous in describing their ideas and practices. The innovative farmer is usually helped by an extension worker or a researcher to write a paper for the workshop. After the paper is presented, the innovation is discussed in detail; the discussion is recorded and a report is prepared which is read to the participants, including the innovative farmer, to make sure the innovation has been properly understood. The participants then decide what further action is needed by research or extension.

The novelty of these workshops is the reversal of roles between extension workers/researchers and farmers. The farmers become the teachers and the other participants learners.

(Ref. Abedin and Chowdury, 1989)

4. GROUP DIAGNOSIS

A New Approach from ICA, Colombia

The Instituto Colomniano Agropecuario (ICA) has developed a method for group diagnosis of farmers' problems. Members of the local community attend group meetings in response to an open invitation, and thus groups form on the basis of self-selection. The meeting dedicates several hours to a brain storming session in small groups, each of which ranks problems in order of importance and assigns a score to them. These scores are used to arrive at a plenary consensus about the problems of most interest to the community.

This approach, constituting a direct input by farmers into the research planning process, has proved highly effective, but it has also been found to mask information about the needs of subgroups. The most important subgroups is women, who are usually under-represented at public meetings in Colombia. Efforts are now being made to form special interest subgroups to carry out the same diagnostic process.

(Ref. Ashby, 1990)

5. ON-FARM RESEARCH

Success and its Modification by ICTA, Guatemala

Founded in 1973, Guatemala's national agricultural research institute, the Instituto de Ciencay Technologia Agricolas (ICTA), was a pioneer of the farming systems approach to research. At first the new institute developed its own strategy for technology transfer, making little or no attempt to link with the government extension service. The assumption, typical of early farming systems research efforts, was that the technology emerging from on-farm trials would be adopted and disseminated directly and spontaneously by farmers.

The assumption proved partly right. Several new crop varieties have spread spontaneously among commercially oriented small-scale farmers in the more favourable low land environments. However, a different approach was needed to reach resource-poor farmers in the more complex and marginal highland areas. To reach these farmers, ICTA had to develop more systematically with the extension service. After several unsuccessful attempts to build these links the two institutions have now launched an integrated research extension programme that appears to be highly effective.

(Ref. Ortiz et al., 1989)

6. RESEARCH – EXTENSION – INTEGRATION

A Success Story from Guatemala

Launched in 1986, Guatemala's integrated research-extension project, involving 72 agricultural extension teams backed up by 20 on-farm researchers, has had a rapid impact on food crop production. It has resulted in an estimated increase in the 1989 harvest of some 16,200 tonnes of staple grains and potatoes, enough to meet the annual basic food requirements of some 26,000 rural families. The project has resulted in about 80,000 resource-poor farmers – approximately 10% of

Guatemala's farm families – adopting new, low-cost technology. The following factors account for the project's success.

Relevant Technology: The project was able to draw on the results of 10 years of research by ICTA, whose client-oriented approach had ensured the development of relevant, low-input technology. In addition, seed production, which is frequently a 'missing task' in the national systems of developing countries (See Section 3), was successfully integrated in the technology development and transfer cycle by relying on the efforts of farmers rather than seed companies.

New model for technology transfer: To launch the project, a new model for technology transfer was introduced to participants in the five regions to be covered. The model stressed the involvement of researchers, extension workers, rural leaders, and farmers as vital links in a single chain, reaching over 2,400 families a year in each region. It served as an important means of fostering a sense of interdependence and a shared set of objectives.

Joint Training: Extension workers were jointly trained with researchers in on-farm research and technology transfer. This enabled them to understand the process through which technology goes before being selected for transfer, helped them to believe in the technology and to 'make it their own' through their knowledge of how it performed and how it should be managed.

Specialization in technology transfer: The joint training led directly to a change in extension strategy. The traditional approach based on adult education and technical assistance was replaced by a new approach with a sharper focus on technology transfer. Extension workers thus became specialists in technology transfer, relinquishing the other tasks for which they had previously been responsible.

Use of rural leaders: Rural leaders acted as a critical link between extension workers and farmers, enabling new technology to reach far greater numbers of farmers than had previously been possible. As well as promoting new technology, rural leaders acted as a channel for feedback to guide the research programme, and helped establish a seed distribution system.

Active Farmer Participation: Farmers also became actively involved, not only in field activities but also in the planning of all research and technology transfer activities. Special consultative groups were formed to ensure participation in planning. The research-extension team presented its proposed work plan to each group for the coming season.

Links at Multiple Levels: Besides joint field activities, the project had three groups at different hierarchical levels which planned and coordinated its work. Links at more than one level are a key feature of successful, integrated systems.

(Ref. Ortiz et al., 1989)

7. STRENGTHENING LINKAGES

Policy Intervention in Zimbabwe

The current strength of on-farm research and its links with extension in Zimbabwe stems from national-level intervention following a radical change in government in 1980. Prior to 1980, the main focus of agricultural policy was the commercial farming sector, which produced most of the internal food crop requirements and export commodities. The government that emerged after independence in 1980 had a strong orientation towards rural development, which included stimulating agriculture in the peasant farming sector.

The on-farm client-oriented research initiative in Zimbabwe was in direct response to this policy shift. Within six years of the implementation of the new policy, nine out of the 17 semi-autonomous research institutes and stations which form part of the national research system had expanded their research agenda to include an on-farm research component. Peasant farmers are now seen as an important client group and most scientists view on-farm research as an effective complementary means for generating technology to meet this group's needs.

(Ref. Avila et al., 1989: Shumba and Fenner, 1989)

8. STRENGTHENING LINKAGES

Policy Intervention in Colombia

In the Narino Highlands in Colombia, the dramatic improvement in output by peasant producers in the 1978-85 period was directly related to the high level of system integration at the time. This integration was facilitated by the national government's decision to increase State involvement in the modernization of peasant agriculture.

The government established the Integrated Rural Development Programme (DRI), gave it the task of strengthening the role of the peasant sector in the national economy, and provided it with considerable resources to achieve this goal. This enabled the DRI to play an active part in coordinating the activities of the core technology development and transfer institutions. As a result, several important linkage mechanisms were developed.

In 1985 the focus of national policy shifted from the highland peasant farming areas to the lowland agricultural frontiers. As a result, DRI budgets have been cut and funding for agricultural technology programmes has been severely reduced. This has contributed to a decline in system integration. Many of the effective linkage mechanisms have disappeared, with collaboration between research and extension reverting to largely informal links. Although agricultural performance is still high in some sectors, it relies upon past, rather than new, achievements.

(Ref. Engel, 1990)

9. LINKAGE MECHANISMS

Impact of Foreign Donor Involvement

Donor involvement in the Philippines had very different effects on the linkage mechanisms created to develop and transfer three types of technologies – corn downy mildew control, seed potato production, and soil improvement and erosion control in hilly lands.

Corn downy mildew control: Foreign donor involvement in this case was limited to the scholarships provided by the United States Agency for International Development (USAID) for foreign training of university professors, and the provision of germplasm by an international agricultural research centre (IARC) to develop resistant varieties. In terms of links within the national system, the effect of IARC involvement was that:

— it contributed to the university's capacity to develop relevant technologies, thus creating favourable condition for links to be strengthened and sustained.

Seed potato production: There was extensive involvement by two foreign agencies in developing and funding production technologies. One worked closely with the Bureau of Plant Industry, a national institution: the other worked closely with the Northern Philippines Root Crops Research and Training Centre. In terms of links, the effect of foreign involvement was that:

— both national institutions established strong but separate links with extension workers and farmers the uncoordinated support provided by the foreign agencies to the national institutions promoted competition rather than collaboration, leading to an almost total breakdown of communication between the institutions: this resulted in task overlaps and duplication of efforts, and attempts made at top government levels to rectify this met with little success.

Soil improvement and erosion control: In this project, an international agency helped to prepare a farming systems development project and financed Cornell University's participation in the project. Cornell worked closely with the local university involved and with the relevant government department. In terms of links, the effect of international involvement was that:

the foreign donor and university developed a good rapport with personnel in the government department and the local university, enabling them to achieve a key project objective – to promote collaboration between the two institutions.

— a Farm Resources Management Unit was set up within the local university, and now serves as a mechanism for sustaining collaboration between the university and the government department in various ways, including staff exchanges.

The third case illustrates the basic conditions necessary to ensure that foreign involvement has a positive impact on links. The agencies worked with all relevant national institutions. They made improved inter-institutional collaboration an explicit goal; and they allocated sufficient time and money to the realization of this goal.

(Ref. Bernardo, 1989)

10 REORGANISATION OF LINKAGES

Failure of a Donor-Driven Reorganisation

Senegal's national research institute, the Institute Senegalais de Recherches Agricoles (ISRA), was established in 1974. Nationalising various programmes that had previously been managed by several French tropical research institutes, the new institute inherited a complex organizational structure and a disciplinary emphasis that made coordination and integration difficult.

In 1975, ISRA prepared a five-year national research plan. In 1982, the institute accepted a major organizational reform proposed by the World Bank in order to implement the plan. Among the changes were the formation of multidisciplinary commodity teams and the transfer of managerial responsibility for research programmes from regional centre heads to the directors of research department. A major component of the reorganisation was the establishment of a new Production Systems Department. This was a large multidisciplinary department with several functions – regional on-farm client-oriented research, research-technology transfer links, and factor research on soil and water management as well as animal traction.

Given the international interest in farming systems research at the time, considerable financial and technical resources quickly flowed into the new department. This made it vulnerable to charges of empire-building from other departmental directors and centre heads, who were seeking to protect their own programmes and positions in an uncertain institutional environment. Competition over resources isolated the Production Systems Department and actually served to seriously undermine links with other research programmes as well as with technology transfer and rural development agencies.

From 1978 to 1989 ISRA had five successive Directors General, each of whom took a different approach to the complex issues raised by the reorganization. This rapid turnover did little to resolve matter. The deepening institutional crisis overflowed into the political area when the government, in response to international pressure to reduce public sector employment, laid off a third of ISRA's staff. Eventually, an uneasy compromise was reached between the former organizational structure and that based on the World Bank-sponsored reform.

Ironically, despite reorganization and donor intervention—perhaps even because of them, ISRA's programmes have continued to operate much as before. The Production Systems Department remains isolated, inter-departmental collaboration in general is still weak, procedures for coordinating research activities at the regional centre are lacking, and research priorities have yet to be revised.

The Moral: When reorganizing, adopt an evolutionary rather than a revolutionary approach. The impetus for change should come from within the institute, not from outside it. Change must take place in a participatory, not a top-down, fashion. When considering adding new units or activities, government and donors must gauge the institute's capacity to absorb them and weigh the costs and benefits of the proposed changes to existing interest groups.

(Ref. Faye and Bingen, 1989)

11. TAILORING LINKAGE STRATEGIES

A Gambian Model

The Gambia's agricultural research service is relatively young and concerned mainly with a PhD. It is heavily dependent upon expatriate staff and external funding. In contrast, the extension service is fairly strong and well established. Staff are generally well qualified, the extension worker-farmer ratio is among the highest in Africa and, although the service relies to some extent on donor support, it receives a greater share of government funding than research. In addition, there is strong involvement by nongovernmental organizations (NGOs) in technology transfer.

In the 1980s, to increase the relevance of research, several attempts were made to introduce on farm client-oriented research into the system. These efforts met with limited success, primarily because they overlooked the need to match donor project objectives with the capacities of the research system. They over stretched the capacity of the research department to develop and monitor on-farm research, while they under utilized the capacity of the extension service and NGOs to play an active and substantial role in these projects.

The Farmer Innovation and Technology Testing programme introduced in 1989, however, stands a greater chance of success because it taps current strengths, rather than making new demands on weaker elements of the system. It is based on using existing farmer groups, formed by extension workers, to conduct verification trials.

Researchers provide information on available technologies; farmer groups, with the assistance of extension workers and NGOs, select those they wish to test; and researchers then assist in trial evaluation.

(Ref. Sompo-Ceesay and Gilbert, 1989)

12. LINKING WITH NGOs

A Bolivian Model

In 1983, in the face of severe resource constraints, the extension programme of the Centro de Investication Agricola Tropical (CIAT) was discontinued and the institution sought to develop links with a number of other groups in Bolivia in order to transfer its technologies to farmers. These groups were termed 'intermediate users', and comprised:

- producers' associations representing both small-and large scale farmers;
- nongovernmental organisations (NGOs) which generally worked with small farmers;
- government extension workers in the field;
- commercial companies.

With the support of the British Tropical Agricultural Mission, CIAT carried out a diagnosis of the activities of technical staff involved in research and technology transfer in these intermediate user groups. It then established various linkage mechanisms with the groups:

- representatives of intermediate users attend annual CIAT meetings to put forward what they consider should be research priorities;
- the groups collaborate in conducting joint on-farm trials, managed and evaluated by CIAT;
- CIAT offers short-term training courses, in the field or non-station, for farmers, NGOs, and others;
- a unit has been established within CIAT to maintain links with intermediate users, partly through the publication of a journal to which all groups contribute articles.
- an information centre has been established to collate information from the intermediate users and distribute it.

This arrangement has succeeded in achieving greater coverage than that achieved by either the traditional extension services or the farming systems approach operating before 1983. It has also succeeded in offering a service in which there are now many more degree-level professionals engaged in technology transfer.

(Ref. Thiele et al., 1989)

13. MERGING RESEARCH AND EXTENSION

A Colombian Case Study

The Instituto Colombiano Agropecuario (ICA) is the country's public-sector national research and extension body, with a mandate to improve performance in many types of crop and livestock production throughout the country. The Coffee Growers Federation is a quasi-private organization with a regional focus, dedicated mainly to the production and marketing of coffee. In ICA, research and extension were merged in 1968; in the coffee federation, the two activities have been together since the federation was founded.

Following the 1968 merger, ICA became a larger, more complex organization, with a broader range of clients, including resource poor farmers. Management had difficulty in focusing on long-term goals or carrying out detailed analysis of specific problems. Conflict between ICA's research and extension groups grew, fueled by overlapping mandates, status differences, and competition for resources. In addition, research became politicized through its association with rural development. The failure to effectively integrate research and extension at ICA may well have led to the decline in institutional performance noted by several observers from the mid-1970s onwards.

In contrast, interaction between researchers and extension workers in the Coffee Growers Federation is more intensive and better coordinated. Unlike ICA, the federation concentrates on a single crop and clientele. Researchers and extension workers have been able to focus more sharply on a narrower range of common concerns. The federation has created a strong institutional culture in which status differences and the competition or resources have been minimized. Lastly, its private status has enabled it to avoid politicization. The overall result has been a high level of performance : coffee yields in Colombia have increased rapidly, and there has been an effective response to coffee rust and coffee bean borer, two major threats to the industry.

(Ref. Kaimowitz, 1989)

14. RESEARCH-EXTENSION LIAISON POSITIONS

Implementation Problems in Zambia

In the early 1980s, Zambia began to use research-extension liaison officers (RELOs) as one of several mechanisms aimed at bridging the gap between on-farm research and extension. Extension professionals are seconded to provincial multidisciplinary adaptive research teams. RELOs have proved very useful for such tasks as bringing researchers and extension workers together to revise crop recommendations, organizing on-farm verification trials of promising technology, coordinating provincial demonstration programmes, and preparing extension materials and newsletters. They have also increased the feedback from extension to research and have begun to influence the direction of research. In so doing, they have helped raise the status of extension workers.

Despite these achievements, there have been several implementation problems. The most important is that RELOs, awkwardly straddled between two departments, have to report administratively to the Provincial Agricultural Officer in extension, but technically to the Chief Agricultural Research Officer.

This has led to ambiguity as to who is responsible for recruitment and performance evaluation. Promotions have also been difficult to obtain since RELOs are integrated within a research team, yet it is their extension supervisors, with little direct knowledge of their performance, who must recommend them for advancement. Ambiguity in job descriptions is a second problem. It has frustrated the recruitment and retention of competent staff and has led to some duplication of tasks with subject-matter specialists.

If these administrative problems can be solved, RELOs could prove to be a very effective link between on-farm research and extension. Familiar with the extension service from which they are seconded, yet committed to the technology developed by research, they are well placed to become product champions, enlisting the support of extension workers in broadening the impact of research.

However, making a success of the RELO experiment will clearly require sustained commitment on the part of senior managers.

(Ref Kean and Singogo, 1990)

15. EXTENSION - RESEARCH LIAISON SERVICE

Nigerian Experience

When responsibility for research in the northern region of Nigeria was transferred from that region's Ministry of Agriculture to Ahmadu Bello University in 1963. The ministry set up a research liaison service to ensure that research continued to respond to the needs identified by the ministry. The service performed well in its early years. Extension specialists, based in research departments, concentrated their efforts on dissemination activities and developed effective research-extension skills.

Between 1969 and 1975, when the ministry was decentralized into six separate state institutions, the liaison service, which was to serve all states, was incorporated into the university's Institute for Agricultural Research. It then became the Extension-Research Liaison Service. As part of the university's system, the service's staff became more professional and independent, and took on more responsibility for adaptive on-farm research. Direct collaboration with researchers diminished and relations became more conflictive, as extension staff began to challenge the relevance of on-station research.

In 1975 the university decided to separate the liaison service from the research institute for two reasons: the service had developed a solid subject matter base of its own, and autonomy would allow it to criticize research without fear of reprisals.

Autonomy proved a mixed blessing, however. It provided the freedom to criticize research and stimulated the development of expertise in generating specialized extension communication materials and training events. But, at the same time, it led to reduced contact with the research institute and an expansion of the roles and activities the service was expected to perform. The service, now named the Agricultural Extension Research Liaison Service, became an implementing, rather than a coordinating agency.

Its limited resources by further expanding the number of client institutes whose needs it was expected to meet, the service is now overextended and underfunded.

The Nigerian experience shows that, for liaison units to be effective, managers must strike a delicate balance they need to restrict both the power and the scope of such a unit in order to prevent the duplication of activities and the dilution of impact. Maintaining this balance is more difficult in large countries with organizationally complex national systems.

(Ref. Ekpere and Idowu, 1990)

16. BUILDING A LIAISON COMMITTEE

A Success Story from Zimbabwe

Shortly after independence, Zimbabwe launched on-farm research in the country's peasant farming areas. The Department of Research and Specialist Services (DR & SS), responsible for research, had no formal links with AGRITEX, the extension service, to cover these areas. Individual stations and institutes within the decentralized research programme of the DR & SS sought ad hoc links with extension. This approach soon led to duplication of activities, inefficient use of resources and overburdening of extension with research support activities. Staff dissatisfaction with this state of affairs triggered the search for better ways of linking research and extension.

The strong linkage mechanism that now exists—the Committee for On-Farm Research and Extension (COFRE)—evolved gradually. The planning meetings that preceded it were too large (40-50 people) and lacked sufficient authority and consensus to make and implement decisions. In contrast, COFRE is a smaller, more manageable group that includes representatives of the directorates and senior management of both DR & SS and AGRITEX. It is a decision-making body, not merely an advisory one. Committee members take their responsibilities seriously, and chairing the committee is considered a major responsibility, expected to take up to 40 per cent of the time of the person responsible. Most importantly, funds have been provided to meet the committee's operating expenses.

COFRE has acted decisively in several key areas. It has increased the efficiency of on-farm research by implementing a system of clustering trials to reduce transport costs, decentralizing the acquisition of inputs by the extension services, and using simpler trial designs. It has played a crucial role in coordinating, planning and priority setting in both departments. A major function of the committee is to screen on-farm research proposals, recommending projects for funding or termination. The clarity and relevance of these proposals is reported to be improving, as researchers become aware that committee support is crucial to the survival of their project. COFRE has now started to exert pressure on the system by requiring that the on-farm research results (not just proposals) be presented to the committee.

COFRE has succeeded in increasing the participation of AGRITEX in on-farm verification trials, in more rapidly translating research results into demonstration projects, and in synthesizing research results into tentative recommendations for several important crops and agroecological zones. It has fostered the growth of a farming systems perspective in AGRITEX, which now includes topics such as tillage systems and moisture conservation in its on-farm trials.

Reasons for COFRE's success: Why has COFRE succeeded when so many committees fail? It is a decision-making body, not just an advisory one. The committee is well represented at the operational level so that staff who have to implement its decisions identify with them and also have a means to influence them. It was initiated by research and extension staff and responds to their felt needs. It has maintained flexibility which facilitates the incorporation of new and innovative ideas.

(Ref. Shumba and Fenner, 1989)

17. REORGANIZING RESEARCH AND EXTENSION

Failure in Costa Rica

The Ministry of Agriculture and Livestock in Costa Rica is responsible for both research and extension. Between 1980 and 1989, the ministry went through numerous reorganizations designed, among other things, to improve the links between research and extension. These changes included: putting research and extension in the same department; revising the extension organogram several times; creating a research-extension planning unit; regionalizing research and extension; adopting the Training and Visit system of extension; reorganizing research into commodity programmes instead of disciplines; and forming commodities programmes involving research and extension.

With reference to maize technologies in the Atlantic Coast region, none of these changes had an important, identifiable impact on either research or extension or on the links between them at the field level. There were few changes in the number of research and extension activities, their topics, or their methodology which could be attributed to the reorganizations. Links between the two activities remained weak. Extension workers and farmers had no effective influence on research planning and there were few mechanisms for making research results available to extension. Adoption of maize technology has continued to be low.

The study concludes that the reorganizations were too frequent. Not enough was done to educate staff about the reasons for the changes or to persuade them of their necessity. There was insufficient attention to field level links and methodologies. Resource limitations were at the root of many of the field-level problems, but these were not addressed by changing the boxes in the organograms.

(Ref. Palmieri, 1989)

18. JOINT PLANNING IN THE FIELD

A Story of 'Group Trek' in Nepal

The difficulties of travel and communications in the rugged terrain of Nepal hinder the coordination and collaboration of field work. To overcome these obstacles, researchers have developed the 'group trek' Researchers from different programmes, departments, and/or stations travel together at regular intervals to a target area where they meet for several days to plan and review their research activities. They diagnose problems and set priorities based on interviews with farmers and local officials and group discussions. Agreement is reached on the appropriate division of responsibilities and a plan for coordinated research is formed on the spot. Researchers organize follow-up meetings to monitor on-going activities and review research results.

The group trek has been easier to use in the two, externally funded, regional research centres in the highlands. Their small size and regionally focused mandate facilitate collaboration. The scientists live together at the station, see themselves as working in an integrated programme, and report to a common director who actively promotes joint exercises. Moreover, target areas are within one day's walk and donor funding ensures adequate per diem.

In contrast, the Farming Systems Research and Development Division, which is a single department in a large national research system, has faced more daunting problems in conducting group treks. It is organizationally and logistically more difficult and costly to bring together scientists from different departments and dispersed research stations and to support a nation-wide network of target areas. Motivation is a second problem. Per diem in the national system do not even cover the actual costs of being in the field. Nevertheless, the group trek, although now conducted less often and for shorter periods, continues to serve as the primary mechanism for linking farming systems research with the station-based research of other divisions and commodity programmes.

(Ref. Kayastha et al., 1989; Mathema and Galt, 1987; Merrill-Sands and McAllister, 1988)

19. INTERNAL REVIEW WORKSHOPS

A Case Study from Bangladesh

After three seasons of on-farm research in the Extension and Research Project in Northwest Bangladesh, researchers decided to hold a workshop to review the previous year's results, diagnose problems, and prepare plans for the following year. They also intended to discuss input and equipment needs and review inter-agency linkages. Researchers, unsure of their results, decided not to involve extension workers because they did not want to lose their confidence. Only an expatriate consultant to the Department of Agricultural Extension and a regional extension director were invited to attend.

The outcome was disappointing. Only a few researchers produced technical reports; most lacked experience in how to analyze data and write reports. What was intended to be a two-day meeting lasted only two hours.

Although initially frustrating, the problems encountered in the workshop led to the organization of training for researchers on data analysis and preparation of technical reports. This training yielded much better results at subsequent meetings. The workshops were also redesigned to strengthen their role as a mechanism for linking on-farm research, on station research, and extension. Extension workers from the districts were invited to participate and were asked to report on their demonstration trials in order to give researchers feedback on how recommended technologies were performing in farmers' fields and how they might be modified. The meetings were also expanded to include representatives from other research institutes active in the region.

Institutionalized as Regional and District Technical Committees, the workshops became the central mechanism in the project for linking on-farm research, on station research, and extension. They gave staff the opportunity to work together to discuss methods and results and to plan research and extension strategies. They also served as an important mechanism for improving the quality of on-farm research and interpretation of results.

The mechanism worked successfully for five years. It proved difficult to sustain and operate effectively, however, when the project expanded from six to ten districts and the research component was transferred from a regional station to the national headquarters. Its effectiveness appears to have declined as many new people became involved who had little orientation to the project, on-farm research methods, or the workshops' objectives and functions. Moreover, at regional level, both the director of extension and the research team leader were personally responsible and accountable for strengthening links. When the project was moved to the national level, no one was assigned equivalent responsibilities for making the links work.

(Ref. Abedin and Chowdury, 1989)

20. RESEARCH -EXTENSION COLLABORATION

An Experiment in Zimbabwe

In 1980, the Agronomy Institute of the Department of Research and Specialist Services (DR & SS) launched a major on-farm trial programme with a view to adapting existing technologies to the particular conditions of the peasant farming areas. The trials were designed by research staff, covered a range of crops and were widely dispersed throughout the country. Because of the scale of the programme, researchers delegated, on a rather informal basis, the routine management of the trials to local extension workers.

The results were disappointing. The researcher in charge of the programme had to travel frequently to supervise the trials, leaving him little time to interact with extension workers and farmers. The extension workers had many other responsibilities and little training in managing on-farm trials. As a result, many trials were lost altogether, and few useful data were fed-back into the research process.

The researchers subsequently reorganised the programme, bringing it fully under their supervision and management. They greatly reduced the number of trials and clustered them in representative sites. Research technicians were then outposted to the sites to manage the trials. The research results have improved markedly with better management. The links with extension workers and farmers have weakened, however, with their significantly decreased involvement in the trials.

(Ref. Avila et al., 1989; Shumba and Fenner, 1989)

21. RESPONSIBILITY FOR RECOMMENDATIONS

Experience of Zambia

In the 1970s, researchers at the central research station near Lusaka drew up and released recommendations. These recommendations reflected only minimal extension worker involvement and did not take sufficient account of the particular agroecological conditions of each province. This strained research-extension relationships. Extension workers considered that much of the research conducted at the research stations was not relevant to resource-poor farming conditions. Researchers tended to blame extension workers for the failure of resource-poor farmers to adopt research recommendations.

In 1978 the Minister of Agriculture introduced several key linkage mechanisms aimed at improving the research extension relationship. These included the establishment of Provincial Research and Extension Committees which, among other things, were intended to give research and extension staff the opportunity to suggest changes in research recommendations. In several provinces, the research-extension liaison officers (RELOs), who are members of on-farm research teams deployed in the provinces, have taken a leading role in bringing researchers and extensionists together to revise recommendations to make them more appropriate for provincial conditions. They have brought them together to resolve contentious issues, and incorporated information from researchers, farmers, and extension workers into provincial crop recommendations, using a farming systems perspective.

(Ref. Kean and Singogo, 1990)

22. USING INFORMAL LINKS

What Contributes to its Success

In the 1978-89 period, the main entities involved in food crop technologies in central-west Cote d'Ivoire were:

- a regional development body;
- a regional research institute;
- an integrated regional development project.

In the first period, a memorandum of agreement between the development body and the research institute provided the basis for formal collaboration, but there were no informal links. The research institute assumed responsibility for conducting on-farm trials and developing improved cultural practices, particularly intercropping. Few new technologies were developed during this period.

In the second period, the integrated development project was launched. The project's research and development (R&D) unit conducted on-farm trials and demonstrations. At the same time, the research institute was intensifying its on-farm activities. There were close personal ties between the leaders of the R&D unit and the on-farm research programme. A memorandum of agreement laid the foundation for formal collaboration between the project and the institute in on-farm activities and the project funded some of the institute's operational costs. Several new varieties were developed and widely adopted.

During the third period, the leader of the R&D unit left. His successor did not have the same personal ties with the on-farm research programme leader. Formal links between the project and the research institute remained, and the project continued to fund the institute's on-farm activities. Some new varieties were developed but none was successfully transferred.

Stability, adequate resource and formal links were features of all three periods. What made the difference in the second period? —strong informal links, developed largely from the friendship between the two leaders and from their strong commitment to collaboration and achieving impact. The recognition of interdependence which emerged as a result of informal collaboration converted 'passive' links into 'active' links In the third period, when informal links broke down, the relationship between the two entities reverted to passive collaboration.

(Ref Eponou, 1990)

23. DEPUTATION OF FIELD STAFF FOR ADAPTIVE RESEARCH

Experience from Zambia

In setting up the Adaptive Research Planning Team (ARPT) programme, no provision was made initially for the secondment of extension workers to field research programmes. However, when the programme organisers began to plan the trials for the first season, they decided that instead of recruiting technicians and posting them to research sites to supervise the day-to-day on-farm research operations, it would be more effective to draw on full-time extension workers already living in the regions. The Extension Branch was willing to second some of its best local agents because extension had been actively involved in planning the ARPT and saw it as a mechanism for generating technology relevant to local conditions.

Known as Trials Assistants, these personnel are responsible for the day-to-day management of on-farm trials, for maintaining contact with farmers, and for organizing field days. They have been effective not only because they received adequate training but also because they are from the areas in which they work, speak the local languages, and understand local farming practices and food preferences.

Trials Assistants are paid by the Extension Branch but supervised by the Research Branch. This had led to some conflicts but, with good communication at the provincial level, such conflicts are usually resolved before they become serious problems.

(Ref. Kean and Singogo, 1990)

24. COMMUNICATION PLANS

It Helps Technology Transfer

In 1981, the communication department of Colombia's national research and extension institute developed a new approach to improve technology transfer to small farmers. The methodology focused on the planning, preparation, and implementation of multimedia extension programmes and provided the means to clearly define the objectives, target groups, and content of these programmes. A key element was the systematic planning of extension events with groups of farmers in a village.

The Communication Plans were designed and implemented jointly by extension field supervisors, subject-matter specialists, and researchers. The involvement of specialists meant that the plans were backed up by a continuous production of high-quality extension materials, including slide shows, booklets, brochures, a newsletter, and later on, a daily radio programme providing news on meetings, demonstrations, and research recommendations. These materials were designed to ensure easy access by peasant producers, women, and children. For example, written materials used peasant vocabulary for technical words.

The Communication Plans significantly increased the number of small farmers reached by the research extension institute.

(Ref. Engel, 1990)

25. USING DIFFERENT LINKS

A Case Study from Colombia

The main agencies concerned with rice technology in the region known as Llanos Orientales in Colombia are the national agricultural institute, a growers' organization, an international research centre (Centro International de Agricultura Tropical – CIAT), agrochemical and seed production companies, and private agronomists. Although all these agencies were involved in the development and transfer of four different types of rice technologies – imported varieties, nationally developed varieties, fertilizer recommendations, and cost-reduction techniques – the links between them differed with each type of technology.

Imported Varieties: The first major technological improvements were based on imported varieties. The growers' organization was closely involved in importing, multiplying and distributing improved varieties. One variety, Bluebonnet-50, was eventually planted in over 90 per cent of the rice area. This success rested entirely on simple trial and error, carried out on a purely commercial basis without any elaborate research efforts. There was minimal coordination between the growers' organization and other agencies.

Nationally Developed Varieties: Despite its success, Bluebonnet was susceptible to local diseases and had rather low yield potential. Thus in 1957, a research programme was created to develop varieties adapted to local conditions. Through a formal agreement, the agricultural institute assumed responsibility for breeding and the growers' organization took on the responsibilities for extension, seed multiplication, and seed distribution. The two agencies shared responsibility for regional variety trials and for determining the agronomic recommendations to accompany new varieties. After the first locally bred variety was released in 1963, coordination grew even stronger through staff secondment, training programmes, and regular joint planning and priority-setting meetings.

With the creation of CIAT in 1967, a triangular relationship developed between CIAT, the research institute, and the growers' organization; CIAT provided support in the early breeding stages and in variety evaluation. Around the same time, the growers' organization exchanged its direct technical assistance role for a coordinating role as growers' needs became more sophisticated and the number of private agronomists increased as a result of government credit programmes. In addition, the growing number of private seed multipliers and distributors provided informal but effective communication links between researchers, private agronomists, and farmers. The clear division of responsibilities and strong links (many of them informal) which developed between these agencies were a major factor in more than twofold increase in rice yields between 1961 and 1984.

Fertilizer Recommendations: The increase in new varieties has been accompanied by a need for fertilizers. Most soil research is conducted by the agricultural institute. However, there is little coordination between this institute, the growers' organization, and the private sector in this respect, or even within the institute's own soil and rice units. The institute relies mainly on publications and seminars to disseminate results, but faces strong competition from private companies, who have far greater resources to promote their products. The lack of coordination has meant that, although fertilizer use has increased, the specific fertilizers used and their methods of application often differ from those recommended by researchers.

Cost-Reduction Technique: Yields from the locally developed varieties have remained high, but production costs have soared. In an effort to deal with this, CIAT, the agricultural institute, and the growers' organization have produced a cost-reduction plan. Among other things, the plan recommends that pesticides should be applied only when critical economic threshold levels have been reached. The three agencies have collaborated closely on disseminating information on reduced pesticide usage, mainly through demonstrations and publications. However, acceptance from growers has been low because there has been little practical support for the plan from the private sector or from private agronomists. The growers' organization is now considering working directly with farmers. Although it stopped providing technical assistance in the 1970s, it has recognized that the institutional arrangements that worked well for transferring new varieties are not adequate for transferring information on pest management.

(Ref. Agudelo and Kaimowitz, 1990)

26. STATUS DIFFERENCES

Conflicts in Colombia

In Colombia, researchers from the agricultural research institute wanted to use staff from the Caqueza development project to help them collect data, but felt that analyzing this information or using it to produce recommendations was purely a research responsibility. Project staff, on the other hand, felt that both groups should participate equally in implementation and analysis with equal recognition. They resented always being regarded as junior partners. This led to several clashes between the two groups.

These conflicts, in addition to their dissatisfaction with the technology researchers were offering, led project staff to undertake their own adaptive research efforts. Researchers, on the other hand, were skeptical about project staff's ability to carry out research-related tasks since they had limited formal training in agronomic research. The researchers complained that the development project's research activities duplicated previous work, suffered from faulty experimental design, and confused demonstration activities with experiments.

(Ref. Zandstra et al., 1979)

27. STATUS DIFFERENCES

How it was Reduced in Guatemala

Over 4,000 rural leaders have been hired by Guatemala's Ministry of Agriculture, which pays them a small salary. They work closely with extension agents and groups of farmers in transferring new technology. They also act as a channel for feedback to guide the research programme and have helped establish a seed distribution system.

Rural leaders are influential and respected members of the community, but the difference is that they are elected by that community with an explicit mandate to act as a link between the government agencies and the people. Care is taken not to remove them from the rural environment, not to change them into trainers or bureaucrats. They divide their time equally between working with their group of farmers and working on their own farm.

(Ref. Ortiz et al., 1989)

28. STATUS DIFFERENCES

How it was Avoided in Tanzania

The National Coconut Development Programme (NCDP), a parastatal, established its own extension component with staff seconded from the ministry. These staff provide technical knowledge and services to local extension workers and directly to farmers.

Status differences have been avoided because the NCDP extension staff, who are specialized in coconut production, have been able to develop technical expertise to complement their knowledge of producers' conditions and problems. This has facilitated successful links with researchers. On the other side, links with local extension workers are strong because the two groups have similar backgrounds and experiences and can work together well in the field. The NCDP staff can translate research results into language the field staff can easily understand.

(Ref. Lupanga, 1989)

29. LEADERSHIP

A Quality to be Earned

On-farm research leaders' responsibilities include developing and guiding strong scientific programmes, building effective field teams, managing arrangements with donor agencies, disseminating research results, and developing appropriate links with other scientists, with farmers, and with technology workers.

Successful on-farm research leaders and managers have:

- On-farm experience;
- Administrative and managerial experience;
- A good working knowledge of all the disciplines involved in the programme;
- Personal commitment and the ability to inspire commitment in others;
- The ability to resolve conflicts and foster collegial relationships;
- A forward-looking approach.

Among the most important of these is the ability to inspire commitment in others, both within the programme itself and in the national system as a whole. A strong leader is needed to defend the scientific credibility of on-farm research and champion its cause. "A consensus in favour of on-farm research among the leader's senior colleagues in the national system is a vital ingredient in the mix of conditions necessary if on farm research is to prove institutionally sustainable. It is the leader's responsibility to build this consensus and commitment

To inspire commitment, a leader must earn the respect of all parties involved. Administrative and scientific competence are therefore essential qualities, and must be backed up by an ability to forecast long term needs and make adequate preparations to meet these needs, for example putting in place effective measures to address potential changes in funds, staffing patterns and availability or programme objectives.

(Ref. Bingen and Poats, 1990)

REFERENCES

Abedin, M. Z. and Chowdury, M. 1989, *Organizational and Managerial Innovations for Extension-Research Linkages in Bangladesh.* The Hague: ISNAR.

Agudelo, L. A. and Kaimowitz, D. K. 1989, *Institutional Linkages for Different Types of Agricultural Technologies: Rice in the Eastern Plains of Colombia. Linkages Discussion Paper No. 1.* The Hague: ISNAR.

Ashby J. 1990. *'Small Farmers' Participation in the Design of Technologies.'* In Altieri, M. A., and Heeht, S. B. Agroecology and Small Farm Development. Boca Raton, Florida: CRC Press.

Biggs, S. D. 1989. *'Resource-Poor Farmer Participation in Design of Technologies.* In Altieri, M. A., and Heeht. S. B. Agroecology and Small Farm Development, Boca Raton, Florida: CRC Press.

Bingen, R. J., and Poats, S. 1990. *'Staff Management Issues in On-Farm Client-Oriented Research: Lessons for Managers. OFCOR Comparative Study Paper No. 5.* The Hague : ISNAR.

Bourgeois, R. 1989. *Promoting Integration through Structural Changes: Making the Link between Agricultural Research and Technology Users.'* ISNAR Working Paper.

Ekpere, J., and Idowu, I. 1990. *'Managing the Links between Research and Technology Transfer : The Case of the Agricultural Extension-Research Liaison Service in Nigeria. Linkages Discussion Paper No. 6.* The Hague: ISNAR.

Engel, J., 1990. *The Impact of Improved Institutional Coordination on Agricultural Performance: The Case of the Narino Highlands in Colombia. Linkages Discussion Paper No. 4.* The Hague: ISNAR.

Eponou, T. 1990. *Informal Linkage Mechanisms and Technology Transfer: The PACO Project in Cote d'Ivoire. Linkages Discussion Paper No. 5.*

Ewell, P. T., 1989. *Linkages between On-Farm Research and Extension in Nine Countries. OFCOR. Comparative Study Paper No. 4.* The Hague: ISNAR.

Faye, J., and Bigen, R. J. 1989. *La Methodologie de la Recherche sur les Systemes de Production comme Instrument de Reorganisation et Reorientation de la Recherche Agricole Senegalaise.* The Hague: ISNAR.

Kaimowitz, D. K. 1989. *Placing Agricultural Research and Technology Transfer in One Organisation: Two Experience from Colombia. Linkages Discussion Paper No. 3.* The Hague: ISNAR.

Kaimowitz, D. K., Snyder, M., and Engel, P. 1989. *A Conceptual Framework for Studying the Links between Agricultural Research and Technology Transfer in Developing Countries. Linkages Theme Paper No. 1.* The Hague: ISNAR.

Kaimowitz, D. K. and Merrill-Sands, D. 1989. *Making the Link between Agricultural Research and Technology Users. Conference Discussion Paper No. 3.* The Hague: ISNAR.

Kayastha, B. N., Mathema, S. B., and Rood, P. 1989. Nepal: *Organization and Management of On-Farm Research in the National Agricultural Research System. OFCOR Country Case Study No. 4.* The Hague: ISNAR.

Kean, S., and Singogo, L. P. 1990. *Bridging the Gap between Research and Extension in Zambia: The Incorporation of Research-Extension Liaison Officers into the Adaptive Research Planning Team. OFCOR Discussion Paper No. 1.* The Hague: ISNAR.

Lupanga, I. J. 1989. *The National Coconut Development Project: Subsystem Draft Report.* Paper presented at workshop on Research and Technology Transfer Linkages, September, 1989, ISNAR, The Hague.

Martinez Nogueira, R. 1989. *The Effect of Changes in State Policy and Organisation on Agricultural Research and Extension Links: A Latin American Perspective, Linkages Theme Paper No. 5.* The Hague: ISNAR.

Merill-Sands, D., and McAllister, J. 1988. *Strengthening the Integration of On-Farm Client Oriented Research and Experiment Station Research in National Agricultural Research System (NARS). Management Lessons from Nine Country Case Studies. OFCOR Comparative Study Paper No. 1.* The Hague: ISNAR.

Ortiz, R. Ruano, S., Juarez, H., Olivet, F., and Meneses, A. 1989. *Closing the Gap between Research and Limited-Resource Farmers: A New Model for Technology Transfer in Guatemala.* The Hague: ISNAR.

Palmieri, V. 1989. *Efectos de los Cambios estructurales en el Ministerio de Agricultura Ganaderia de Costa Rica Sobre la Relacion entre Investigaciony Transferencia de Technologia en Maiz.* The Hague: ISNAR.

Roling, N. 1989. *Why Farmers Matter: The role of User Participation in Technology Development and Delivery.* The Hague: ISNAR.

Seegers, S., and Kaimowitz., D. K. 1989. *Relations between Agricultural Researchers and Extension Workers: The Survey Evidence. Linkages Discussion Paper No. 2.* The Hague: ISNAR.

Shumba, E., and Fenner, R. 1989. *Linking Research and Extension through On-Farm Research and Demonstrations: The Zimbabwe Experience.* The Hague: ISNAR.

Sompo-Ceesau, M. S., and Gilbert, E. 1989. *Special Issues for Small Countries in Making the Link between Agricultural Research and Technology Users: The Case of the Gambia.* The Hague: ISNAR.

Zandstra, H., Swanberg, K., Zulberti, C., and Nestel, B. 1979. Caqueza: *Living Rural Development. Ottowa:* IDRC.

ANNEXURE IV

ILLUSTRATIVE SET OF ISSUES FOR FACILITATING DISCUSSIONS IN THE NATIONAL WORKSHOP HELD ON FEBRUARY 1 AND 2, 1993 AT MANAGE, HYDERABAD

It has been generally accepted both in cases of extension and research that there is considerable scope to improve quality and management and cost effectiveness. Some measures to address these have already been identified and are being implemented to varying degrees. To carry forward this process, a study has been commissioned by MANAGE with the following objectives:

Objectives

The first objective of the study is to develop one or a number of options on how stronger farmer participation can be introduced into extension and research management in India and how extension workers and research scientists can be made more accountable to the client groups. The second objective is to recommend how and where the proposed options for introducing farmer participation can best be tested on what scale, and how they may be rigorously and, if possible, empirically evaluated.

Issues

I. *Institutional Innovation*

1. In the Indian scenario, can public sector extension and research services be reoriented in a manner so as to make them more farmer driven-accountable to and controlled by farmers?
2. What institutional innovations, financial changes, methodological improvements and other related modifications in current planning and management of extension and research are

required to support their reorientation?

3. What changes are required at the policy level, organisational structure, standard operating procedures and leadership style to encourage farmer participation in design and implementation of agricultural development activities for being responsive to local needs?

II. *Role of Extension Services*

4. What is the nature of services that are required to be delivered through the extension system in different agricultural situations?
5. Should extension services be broadbased to include all landbased activities like livestock development, agro-forestry, agro-based cottage industry, horticulture etc.?
6. How can extension services be integrated into overall agricultural development?
7. Is there a need for situation specific agricultural extension and research organisation to attain overall agricultural goals? How can they be fostered?

III. *Non-Governmental Initiatives*

8. How can NGOs be involved in a greater way in extension and research services? Should they compete, substitute or complement public extension and research services?
9. What is the scope for privatization of extension and research services? How can cost sharing models of agricultural extension be developed? How can greater contracting of agricultural extension service to the private sector be encouraged?

IV. *Reward or Incentive Systems*

10. How can the morale and motivation of the extension workers be increased so as to enable them to more effectively serve the farmer?
11. Would performance linked rewards and incentive system enable extension workers to participate with farmers more effectively. How can such performance be monitored and assessed?

V. *Linkages*

12. What should be the methodology to increase the effectiveness of research-extension linkages, particularly at the grass-root level? What are the institutional arrangements that are required?
13. How can feedback mechanisms from extension workers to the extension and research hierarchies be strengthened?

VI. *Methodologies*

14. How can real and effective farmer participation be introduced into planning and management of extension and research?
15. How do we ensure that biophysical and socio-economic conditions influencing farmers' decisions on enterprise mix get reflected in research prioritization process?
16. When we talk of farmers' participation, are we talking of functional participation or fundamental participation? Whether farmers' involvement is to be confined to implementation of programmes or should it also extend to designing programmes?
17. How to make extension services gender sensitive? Do we need to develop newer methodologies?

The discussions on the aforesaid issues should ultimately lead to:

(a) identification of different models of successful agricultural extension-research systems in operation in the country, be they in public, private, co-operative or NGO sectors, for indepth study by the consultants;

(b) identification of a reasonable number of distinct agricultural situations in the country for which appropriate extension models are developed by the core groups of consultants; and

(c) formulation of a set of questions to be posed by the consultants to a cross-section of persons involved in agricultural extension and research ranging from client groups, implementors, planners, scientists etc. to facilitate semi-structured interviews.

ANNEXURE V

PROCEEDINGS OF NATIONAL WORKSHOP HELD ON FEBRUARY 1 AND 2, 1993 AT MANAGE, HYDERABAD

At the outset Shri S. K. Arora, Director, MANAGE welcomed the participants. He outlined the objectives of the workshop and highlighted the issues to be discussed at the workshop.

Mrs. Dayal, Joint Secretary, Extension, Ministry of Agriculture, Government of India dwelt at length on the background for commissioning this study and pointed out the need for closer interaction between research system, extension system and farmer. She stated that the overwhelming concern was how to make the present extension and research system farmer driven and responsive. There are a number of models in the country which are in operation using farmer participatory approach. An indepth study of these models may help in formulating a strategy to make research and extension systems farmer participatory, farmer oriented and farmer accountable.

With these introductory remarks the discussions centered around the issues framed for the workshop.

I. Institutional Innovations

The participants were of the view that there is strong need for reorientation of public sector extension and research systems. The lessons available from various NGO models, private and corporate extension systems can help in restructuring the present extension and research systems to become more client driven. Participants cited examples of MYRADA in Gulbarga, AKRSP in Gujarat, BAIF, KRIBHCO Project in Dahod, Ram Krishna Mission in Narendrapur, GROWFED in Gujarat, Sadguru Seva in Dahod. In the Government sector, it was pointed out that the National Watershed Project, Oilseed Technology Mission are carrying on extension work on project mode. In the private sector, fertilizer and pesticides companies are conducting research and extension e.g., ITC, MICO, RCF, HCF, IFFCO, KRIBHCO, Coffee, Tobacco and Tea Boards are also undertaking research and extension activities to serve their clientele.

It was suggested that indepth study of a few of these models may help the consultants in identifying lessons which can be used to effect changes in orientation, policy, planning and financial management, methodology of implementing programmes.

II. Role of Extension Services

The role of T&V system of extension was discussed in great detail. The participants outlined weaknesses in the system that it is focussed on narrow crop based approach being uniformly followed throughout the country with a rigid schedule. It was felt that uniform model of T&V system is not appropriate for different agricultural situations e.g., Hill areas, Command Areas, Arid Zone, High Rainfall Areas, Coastal Belt, Rainfed Areas. The needs of farmers in these areas are quite different from each other. The extension system therefore needs to adopt different approaches to effectively cater to farmers requirement.

Emergence of second generation problems in assured irrigated areas requires extension and research systems to focus their attention to sustainability issues. This calls for a new extension strategy focussing on institutional building, human resource development and group mobilization.

III. Linkages

Various mechanisms of linkages (e.g. Coordination Committees at State and District level, monthly workshops, zonal councils) between Extension and Research were analyzed. The need to make them more effective to facilitate flow of information from research to extension and feedback to research was emphasized. To strengthen linkages at the grass-root level, joint visits of scientists, and extension personnel, innovative farmers workshops, exposure visit of farmers to research station etc. were suggested.

IV. Methodologies

To make research, farmer's problem oriented, it was emphasized that farmers be involved in the process of identification of problems through

PRA techniques. It was also suggested that scientists, extensionists and farmers jointly decide the priorities for research based on the magnitude of the problem. The visits of scientists to have more interaction with farmers in their own situation was also highlighted to understand farmers perception of his needs.

Non Governmental Initiatives

NGO's significant contribution in poverty alleviation and development activities were highlighted by the participants. NGO's like MYRADA, AKRSP, BAIF, Deccan Development Society, ASW have demonstrated potential for closer collaboration between NGOs and the public agricultural research and extension systems.

Participants pointed out innovative and relatively successful approaches like participatory methods of diagnostic and farming systems approach, adopted by Ram Krishna Mission, MYRADA, AWS.

Participants emphasized need for closer cooperation between extension and research systems and NGO's. NGO's role should be viewed as complementary and supplementary to public sector extension and research systems and not as competitive. NGO's experience in innovative methodologies could be profitably used to make research and extension system farmer participatory.

In the end, it was decided that the consultants visit and carry out indepth study of the following NGO's models:

Gujarat	AKRSP, KRIBHCO, GROWFED, BAIF. Sadguru
A. P.	ITC, Agro-Business Division – Sunflower Seed Project, KVK, Guddipalli
Karnataka	MYRADA watershed project

In regard to identification of agricultural situations in the country for which test models are to be developed, it was suggested that ICAR's agro climatic zones be taken as a basis. However, analysis of agricultural situations and differential agricultural extension and research needs of various regions led the consultants to classify the country into the following generic groups:

(1) Rainfed areas
(2) Irrigated areas
(3) High value commercial agriculture region

ANNEXURE VI

REPORT ON FARMER PARTICIPATORY STUDY CONSULTANCY FIELD VISITS TO FARMER PARTICIPATORY PROJECTS

K. C. JOHN, C. S. RAJAN, D. K. SHARMA AND CHARANJIT SINGH

Context

India has experimented with a veritable mix of institutional arrangements to enhance the responsiveness of technological generation and transfer to farmers' needs. More often, a top-down Transfer of Technology (TOT) mode of agricultural research and extension is in operation. Of late, however, there has been a spurt in farmer participatory technology development and dissemination, largely in the voluntary and Non-Government Organisations, but also in a few public sector institutions.

The institutional arrangements, currently operating in the sphere of technology generation and transfer can be broadly classified into the following group:

1.0 Public Sector Initiatives
2.0 Cooperative Sector Initiatives
3.0 Growers' Association
4.0 Private Sector Initiatives
5.0 NGO Initiatives.

1.0 Public Sector Initiatives

1.1 Training & Visit System
1.2 Technology Mission e.g. Oilseeds
1.3 Special Crop Production Programme e.g. Special Rice Production Programme, National Oilseeds Development Programme, National Pulse Production Programme.

1.4 National Agricultural Research System comprising a network of ICAR Central Institutes, National Centres and SAUs.

The Extension Research Linkage is provided through an elaborate arrangement at National, State, Zonal and Sub-zonal level. However, its effective implementation is ensured at Zonal Research Stations through Zonal Research Extension Advisory Committee (ZREAC), monthly and fortnightly meetings and joint diagnostic visits. More recently, ICAR's Division of Agricultural Extension has made an attempt to bridge existing institutional gap by strengthening KVKs at the district level.

A network of State Agricultural Universities in Eastern India has been experimenting with farmer participatory farming systems research extension (FSRE) approaches to generate location specific technologies since 1985. This programme is now known as Eastern India Farming Systems Research/Extension Network.

1.5 *Commodity Boards:* Commerce Ministry has established several Boards to cater to the research, extension, promotion, credit and marketing needs of export oriented crops such as tea, coffee, rubber and cardamom.

2.0 Cooperative Sector Initiatives

2.1 *National Dairy Development Board*

Success of milk cooperatives in Kheda, Gujarat led to establishment of the National Dairy Development Board (NDDB). It embraces a systems approach in organising its activities related to farmers' organisation, research and extension, procurement, processing and marketing. NDDB has diversified its operation to oilseeds and horticultural crops.

2.2 *Fishermen's Cooperative in Kerala*

Similarly, fishermen in Kerala coast have organised a cooperative venture to look after their organisation, research and extension, procurement, infrastructural servicing, processing and marketing needs of marine fishery.

2.3 *Crop Related Cooperatives*

Cotton, Sugarcane, Potato growers' cooperatives have generally tended to assume trading and warehousing responsibilities. Though in Maharashtra Şugarcane Cooperatives have aggressively taken on a systems approach in organising their activities.

3.0 Growers' Association

3.1 *UPASI*: Planters' (Tea, Coffee, Rubber and Spices) of Southern India formed an association known as the United Planters' Association of Southern India (UPASI) in 1893. UPASI' affiliates include the State Planters' Associations of Tamilnadu, Kerala and Karnataka, and the many district level planters' associations in these States. It is financed by an area-based cess on its members. Besides representing growers' interests in national and international policy, UPASI is also engaged in research, information exchange, training and estate supplies. UPASI received an ICAR support to establish a KVK in 1982.

3.2 *Grape Growers' Association in Maharashtra*: Grape Growers in Maharashtra have formed an apex organisation and have adopted a systems approach in structuring its activities.

3.3 *Indian Maize Development Association in Haryana and Karnataka:* Bharat Starch Company in Haryana and Congill India in Karnataka has established Maize Growers' Association recently.

4.0 Private Sector Initiatives

4.1 *Seed Industry*

Several leading companies in the seed industry have established research and development units. They also have an extensive network of contract growers.

4.2 *Agro-Chemical Industry*

Similarly many of the agro-chemical companies carry out both in-house R&D and sponsor research on their areas of interest. They have a well-laid out network of extension and promotion outreach centres. Recently, Hindustan Fertilizer Corporation and KRIBHCO have embarked on a farmer-participatory Rainfed Farming Project in Eastern tribal plateau (Bihar, Orissa and West Bengal) and Western India Tribal regions (Gujarat and Rajasthan and Madhya Pradesh) respectively.

5.0 NGO Initiatives

NGOs have known to be involved in rural and social development. Several studies and reports have now highlighted their role in adaptive research and extension. Bharatiya Agro-Industries Foundation (BAIF) in somewhat "top-down" mode made a significant impact on research and development in livestock production, health and nutrition. Of late, BAIF has embraced bottom-up approach of extension and research.

PRADAN's work has concentrated on innovative income - generating opportunities both on-farm (Mushroom and raw silk production) and off-farm (improved processing of hides and skins : improved handloom designs); Special credit schemes for women and scaling down of technologies. Simultaneously, it has also sought to strengthen local membership organisations capable of carrying forward the programme in a sustainable fashion.

MYRADA has very effectively addressed the issues of innovations in the methodologies of diagnosis and NGO-GO collaboration on Watershed Development. It is also collaborating with International Institute of Rural Reconstruction, Philippines in experimenting with horizontal diffusion and production of technology dissemination kits.

AKRSP besides innovating methodologies of training and establishing farmers' organisations, has initiated farmer participatory resource planning and implementation of watershed, Joint Forest Management, efficient use of water resource and an action research on groundwater management.

Ramakrishna Mission, Narendrapur capitalised on existing rural youth clubs in innovating and systematizing methodologies of agroecosystem analysis and participatory research on whole farm approach to augment the livelihoods of rural poor. Action for World Solidarity and Deccan Development Society have experimented with the role of a match maker in linking grassroots organisations and public sector research and extension services in blending informal and formal research to control Red-headed hairy caterpillar of castor.

Society for Hill Resources Management School has evolved a unique self-revolving and self-sustaining research and development methodology to regenerate degraded wastelands while ensuring equitable distribution of benefits and improved livelihood sustenance of tribals in eastern plateau region.

Auroville Greenwork Resources Centre has successfully

experimented with the principles and practice of sustainable resource management within the existing agro-ecological and socio-economic resources of local communities through fostering inter-institutional collaboration.

Objectives

As a part of farmer participatory study commissioned by Government of India through MANAGE, it is proposed that the team of consultants visit some of the farmer-participatory projects in Andhra Pradesh, Karnataka and Gujarat. The purpose of visits is to learn essential elements of farmer participatory approaches and examine feasibility of its replicability. More specifically, the objectives of field visit are:

1. To understand how farmer participatory approaches were initiated and integrated into organisational (Public/Private/Cooperative/NGO/Grower's Association) activities.
2. To identify common problem, in implementing participatory approaches to technology generation and transfer.
3. To diagnose institutional factors that led to emergence of such problems.
4. To discern institutional or individual/group responses to overcome barriers to client participation.
5. To distill lessons for successful management of linkage mechanisms: Research-Extension; NGO-GO, Cooperative public research and extension, private-NGO-Research extension.

Issues for Exploration

1. Historical organisational profile and genesis of farmer participatory project

    — Donor-driven
    — Professional-led
    — Client-pulled
    — Public System-demanded

2. Bottlenecks to participation and countervailing mechanisms

    — Socio-cultural

— Administrative
— Methodological/conceptual
— Resource scarcity

3. Methodological Issues of Farmer Participation

— Institution Building
— Diagnosis
— Design of Project/Action Plan
— Participatory Planning
— Resource Mobilisation
— Implementation
— Monitoring and Evaluation
— Adoption/Rejection/Modification
— Feedback

4. Management of Linkage Mechanisms

— Policy Issues
— Domain Consensus
— Resource Factors
— Technical Factors
— Planning and Review Process
— Collaborative and Independent Professional Activities
— Resource Allocation Procedures
— Information Exchange and Communication Devices
— Performance Assessment, Rewards and Incentives
— Resolution of Conflict
— Staff Skill Upgradation.

Methodology

Since the main purpose of the visit is to distill lessons from several farmer participatory initiatives, a learning mode would be embraced by the team of consultants. Open ended interviews, group discussions, scanning of published materials, visits to project sites, interactions with clients of both genders and probe questioning would be employed to generate information as well as access key participate analysis.

Itinerary

GROUP I: Dr. Charanjit Singh and Dr. K. C. John

February 8-10, 1993	Private Sector Research and Extension Programme: ITC Agro-Business Division, Hyderabad. Visit ITC's Sunflower Seed Business.
February 11-12, 1993	Public Sector Research Extension Linkages. Visit Krishi Vigyan Kendra, Guddipalli, Nalgonda Dist.
February 13-14, 1993	NGO-GO Collaboration Visit SDC sponsored Watershed Development Collaboration between MYRADA and Dryland Development Board (DLDB), Karnataka, Gulbarga.
February 15-16, 1993	NGO-Public Research Extension Collaboration Visit AWS/DDS-led inter-institutional Project on control of Red hairy caterpillar of castor, Warangal.

GROUP-II: Dr. D. K. Sharma and Mr. C. S. Rajan

February 8-10, 1993	Private Sector Research and Extension Programme: Visit to BAIF's Padhari Project in Ahmedabad District. Cooperative Sector - Integrated Agriculture Service Project: Visit to GROWFED Project area in Mehsana District.
February 11-13, 1993	Private Sector promoted Lift Irrigation Project: Visit to Sadguru Water Development Foundation Project area in Dahod District. – Public Sector promoted research and extension project: visit to KRIBHCO rainfed project area in Dahod District.
February 14-15, 1993	– NGO-GO Collaboration: Visit to AKRSP Project area in Netrang. Bharuch District.

CHECK LIST OF QUESTIONS TO BE ADDRESSED TO THE ORGANISATIONS EXPERIMENTING WITH FARMER PARTICIPATORY APPROACHES

I. *Organisational Profile*

(1) Objectives

(2) Scope and scale of operations:

— Sectors in which operating
— Manpower

(3) Progress of activities

(4) Client Group and Client segmentation criteria:

— Gender
— Income status
— Social Groups

(5) Organisational Structure:

(a) Whether hierarchical or flat?
(b) Span of Control
(c) Division of task(s).

II. *Participatory Approaches*

(1) How was it started—genesis? Source of ideas—whether donor-driven or professional-led or client-pulled?

(2) What is the manner in which the project was originally conceived?

(3) What were the difficulties encountered in implementing the project as originally conceived?

(4) Modification made, if any, subsequently.

(5) What is the current status of the project in terms of the participatory processes?

III. *Methodological Issues of Farmer Participation*

(1) What are the mechanisms that have been evolved to foster farmer participation?

(2) What are the steps involved in identifying the problems?
(3) How was the project designed? Role of client?
(4) Community participation in resource mobilization—How is it being done?
(5) Role of community in implementation, particularly in operation and maintenance? Whether community has a stake in asset creation? What are the agreements of sharing of costs and benefits? Whether there are incentives/penalties for violation?
(6) What are the M&E mechanisms?
(7) What are the conflict resolution mechanisms?
(8) Whether systems have been evolved for facilitating adoptions, rejection or modification of the project in the light of feedback?

IV. *Linkages*

(1) What are the collaborative (mechanisms) institutions available for research extension linkages within the organisation and without?
(2) Whether these mechanisms are formal or informal?
(3) What are the problems involved in drawing or obtaining technical support?
(4) What are the factors that influence the NGOs entering into collaboration with public/private sector systems?

— Cost sharing
— Sharing of ideas
— Planning and review of programmes
— Resource allocation
— Information exchange and communication devices
— Conflict resolution
— Staff skill upgradations.

BAIF

The BAIF is a registered Trust implementing integrated development programmes, starting with large scale livestock development activities for more than two decades. The main target groups are economically weaker section in rural areas. The activities are implemented with support from Rural Development Departments of five States (Gujarat,

Rajasthan, Maharashtra, Uttar Pradesh, Karnataka). In 2-3 States the activities have become more integrated, with introduction of agro forestry, soil conservation, horticulture, vegetable production, water resource development etc. The BAIF is also recognised as a research association by the ICAR and various State governments are involved in implementing applied research programmes. It strongly feels that development activities need to be supported with research, in order to make development more effective.

The BAIF is operating a large number of programmes through network of centres in Gujarat while in most of these centres activities are initiated through development programmes. It has enabled BAIF to form a base in rural areas and establishing rapport with a large number of families. This base has been used for studying, farming systems, traditional methods, social and tribal groups.

At the apex level Board of Trustees, Presidents Council and Central Management Council have been set up to plan, lay down policies, organize, coordinate the activities. An advisory council of scientists has been constituted to provide technological support and guidance.

Project approach has been adopted to implement various programmes. Professionals have been employed to plan, organize and implement the programmes by involving local communities through participatory approaches.

PADHAR PROJECT

Padhar Project was conceived with the object of settlement of tribal families in Devadthal. The Tribal Development Department of Gujarat Government initiated the project in Ahmedabad district in 1984 by allotting 4 acres of land to each tribal family. However no progress could be achieved in relation to development of land, housing etc. BAIF had established a 400 acres farm in Village Camp. At this farm soil salinity was a major problem.

With gypsum application and plantation of salt tolerant trees, crops etc. considerable progress was achieved. Over a period of years the farm became viable. Besides paddy cultivation, the farm has a good dairy of cross breed animals, fruit trees (ber), fodder and fuel trees.

Families of Padhar community were working as labourers on the farm and were actively participating in all types of farm developmental work. BAIF farm staff therefore had a good rapport with these families.

The Tribal Department approached BAIF to undertake settlement of Padhar community in Devathal area. Members of Padhar Community working on the farm were also keen that BAIF staff help them in their settlement.

In 1986 BAIF undertook to implement the project of settlement of Padhar community in Devathal. The project outlined the following objectives:

1. Settlement of Padhar tribal families on land allotted to them in Devathal.
2. To help them in undertaking such enterprises which may provide gainful employment throughout the year so that they are not required to migrate to cities for seeking employment.
3. To improve their quality of life.

Target Group

110 Families were identified for settlement.

Funds for the project came from Tribal Department.

Structure: BAIF appointed an agriculture graduate to work with the community. He was given back up support by farm agronomist, horticulturist, animal scientist and other staff located at Ahmedabad project office.

Approach

Base line information regarding agro ecological aspects, traditional production systems, cultural background, awareness, views, perception of the tribal families, decision making process etc. were collected and analysed to get an overview of the situation. To identify the specific problems (Physical, Bio-social, technological), the following was adopted:

1. Indepth discussions with the community members.
2. Self observation.
3. Discussions with the officers of the tribal department.
4. Studies, reports on Padhar Community as reference material.
5. Discussions with the BAIF farm and other staff based in Ahmedabad.

As a result of discussions, observations and interaction with various groups the following major problems were identified:

1. Padhar Community had large scale unemployment. Fish catching was the main occupation which provides them partial employment. Hence it led to migration to Ahmedabad and nearby towns in search of jobs.
2. Land remains flooded with rain water for 4-5 months.
3. Soil is saline, mono cropped, paddy gives poor yield.
4. They do not have enough cattle population to meet their farming requirements, poor investment capacity.
5. Housing is extremely poor.

BAIF extension worker with support of other specialists assisted the community to identify activities which need to be undertaken. The community decided to undertake: (a) bunding of each individual holding; (b) improvement of saline soil through application of gypsum; (c) Adoption of improved methods of paddy cultivation; (d) construction of pond for fish culture; (e) construction of houses for each family.

The community was actively involved in planning, organizing and execution of the various activities. It was decided that though Tribal Department is to bear the entire cost of different activities, the community will provide manual labour and in no case any of the activities will be executed through outside labour or on contract basis. It was also decided by the community to share the cost of each activity in form of labour and material e.g., for housing 25 per cent, bunding of fields 15 per cent and construction of pond 20 per cent.

Crop farming will be an individual family activity. Fish farming shall be a cooperative venture.

Implementation: In the first phase bunds were constructed on each individual holding. Paddy cultivation was undertaken in monsoon season to utilize the rain water. Farmers were given training in application of gypsum, raising paddy nursery, transplanting, bunding of fields, selection of salt tolerant varieties, application of fertilizers, plant protection measures etc. on the BAIF farm. Extension worker also provided guidance by constantly visiting their fields.

Raising of crops in rabi on residual moisture was also planned. The community undertook construction of 80 acre pond for fish farming.

Fish farming was identified as a priority enterprise. A cooperative society was formed. All 110 families became members of the society. They worked whole heartedly on construction of pond and raising of fishes.

Two members were sent for training in fish farming. The society raised fishes worth Rs. 55000 last year. A bore well was dug to supplement water requirement for fish farming as well as for providing irrigation to crops to be cultivated in rabi i.e., post monsoon.

Low cost housing for all 110 families have been constructed. The housing is on a common piece of land having a single room accommodation with front and rear verandah. A block of 10 such quarters have been constructed to suit the requirement of the community. Families preference was for living in close proximity rather than in dispersed homesteads. The entire community actively participated in construction of houses. Under the guidance of the project staff cement blocks were manufactured on the spot which were used in construction of walls. Tiles, rafter etc., doors and windows were purchased from outside.

Linkages

Project staff has established very good rapport within the community as well as developed meaningful linkages with officers of the Tribal Department and Fisheries Department, who are providing not only funds but also technical support. The interaction with Agriculture Department is totally missing as VEW rarely visits the settlement.

Lessons to be Drawn

1. The project is based on the felt needs of the beneficiaries. Padhar tribes had a strong desire to lead a settled life. High motivation was prevailing in them to improve their way of life. Through this project they could realize their desire for a sustained living.
2. The Project staff especially Extension worker was also highly motivated and committed to project philosophy and goals.
3. BAIF farm provided a suitable back drop for establishing high degree of rapport between extension worker and the community.
4. Participatory approach to identifying problems, needs and also

solutions helped in smooth implementation of the project.

5. Empirical participation in planning, organizing and execution of various activities helped in attainment of goals.
6. Community's share in cost of each activity promoted belongingness and helped in building high morale among community members.
7. Physical, mental and social involvement of each member has instilled confidence in their own capacity and ability.
8. If external agency acts as a facilitator of initial resources, technology, group action, the project can become self propelling e.g., fisheries members are now planning to raise their own seed to meet their requirement and also to market the surplus seed. With bore well water they are now planning for fodder production and dairy farming.
9. Women folk although do not participate in decision making but could still influence the process of decision making e.g., women farm labourers were instrumental in getting BAIF involved in the project.

GROFED

(Gujarat Cooperative Oilseeds Growers Federation Limited)

A Case Study of Farmers' Participatory Approach

The NDDB at the instance of Government of India launched the project for 'Restructuring of Production, Processing and Marketing of Oilseeds and Vegetable Oil' under which GROFED was created in June 1979. The basic objectives of the project were to bring about production enhancement and build in self sufficiency in edible oil requirement in the country. These objectives were to be met by setting up a three-tier cooperative infrastructure comprising oilseeds growers' societies for a village or group of villages, affiliated to Regional Co-operative Oilseeds' Growers Unions covering two or more districts, and an apex body to which the Regional Unions are affiliated.

So far around 1500 village level societies and seven regional unions have been formed in 14 districts of Gujarat. The present membership of the societies is around 2,00,000 farmers.

The GROFED provides integrated services to farmer members involving research, extension, organisational development, input management, market support and post-harvest management. Thus it caters to farmers needs right from production of oilseeds to procurement to processing and finally marketing.

The point of contact between the farmers and GROFED is the village level Co-operative Society which has an elected body with a President, a select number of Executive Committee members and a paid Secretary. Infrastructural support for the Society is provided by the GROFED and technical guidance by the field officers of the Regional Unions. The functions of these Societies are to provide extension services to farmer members, supply inputs like seeds, fertilisers etc., procure oilseeds and market edible oils.

The Regional Unions have a District Officer for each district, and Area Officers for every Taluk and several Procurement Assistants, depending upon the number of Societies. For our study we chose Mehsana Regional Office which comprises the District of Mehsana to the west of Ahmedabad. This district comprises four Area Offices, the brief particulars of which are as follows:

Scale of Operations	Name of the Area Office			
	Mehsana	Vijapur	Visnagar	Patan
No. of Societies	42	39	62	72
No. of Villages	73	68	143	96
No. of Members:				
— Male	7731	7631	12444	8145
— Female	150	100	431	220
Total	7881	7731	12875	8365
Total Oilseed Area	10423 ha	7162 ha	19839 ha	21356 ha
No. of Procurement Assistants	3	4	5	4

Scale of Operations	Name of the Area Office			
	Mehsana	Vijapur	Visnagar	Patan
Avg. No. of Societies per Procurement Asst.	14	8	12	18
Avg. No. of villages per Procurement Asst.	24	17	28	24
Avg. No. of members per Procurement Asst.	2600	2000	2600	2000

Of the aforesaid Area Officers, we visited two Societies in Mehsana Area Office, viz., Piludara Society and Lakhvad Society, the brief particulars of which are as follows:

	Piludara	Lakhvad
No. of Villages	1	1
Population	6000	3500
No. of farmers	750	250
No. of Member farmers	346	110
Total Oilseed area	700 ha.	160 ha.
Oilseed Area of Member farmers	600 ha.	80 ha.

Both the Societies are single villages societies, connected by metalled road with Bank, Post Office, School, Milk Union, PACs and STD. We met the President, Secretary and some members of both Societies as well as the Procurement Assistants. We were accompanied by the Dy. Manager, GROFED, Mr. M. D. Patel. From our discussions with the above persons we realised that while Piludara was a model of a successful society having its own Godown, an active extension and input service, the Lakhvad society was riddled with dissensions amongst the members, between the President and the Secretary, and consequent frustration and disillusionment.

As has already been stated earlier, all Societies have been formed through assistance provided by NDDB at the instance of Government of India; members are required to deposit Rs. 11 as membership fee. Although both these Societies are single village societies, yet their coverage extends to less than 50 per cent of the farmers of the village and in the case of Lakhvad to only 50 per cent of the total oilseed area.

It may be kept in mind that these being Oilseed Growers Societies all other crops are excluded from their ambit. In the larger context of Gujarat we learnt that membership does not exceed 40 per cent of the farmers in the areas covered by the Societies. As a result, although the total Oilseed area in the State exceeds 23 lac ha., the areas covered by the Societies does not exceed 9 lac ha. Thus, one incontrovertible conclusion which emerged was that even in terms of quantitative coverage, GROFED does not account for more than 40 per cent of the oilseed area of the State.

Next, when we come to the quality of services rendered, we found that the Societies largely concerned themselves with procurement and, to an extent, marketing of edible oils. Extension services were being provided through the Procurement Assistants who, often was a B.Sc. (Agrl.) and, occasionally a B. Com. with training in Agriculture.

Inputs were being supplied either through the Oilseeds Societies or the Procurement Assistants or both. The Procurement Assistants helped in aggregating demand, coveying the same to GROFED and arranging supplies to the Societies, who, in turn, ensured availability to the farmers.

We then probed the officials and farmers about the Research Extension linkages of GROFED. We found that GROFED maintains research farms but these were more in the nature of adaptive trial centres where, in addition, seed multi-application programme was being undertaken for supply of seeds to the Societies. The scientists on these farmers maintained liaison with the scientists of the Gujarat Agriculture University and also provided training to the field functionaries of GROFED and the Regional Unions. Thus, barring this, there does not appear to exist any other mechanism of linkage between Research and Extension at the level of GROFED.

At the level of Regional Unions, however, linkages between Research and Extension are maintained through seasonal crop seminars. These are organised jointly by GROFED, the University, Department of Agriculture and the farmers. These seasonal seminars appear to be useful and provide a forum for a lot of interaction amongst all agencies.

At the Area level, the mechanism of monthly meeting is being used to disseminate technological information from the technical officers of GROFED to the field officers. The Area Officers also try and maintain informal liaison with the scientists of the nearby Research Station of the University.

On the whole, it was observed that Research-Extension linkages are weak, not formalised and left largely to the initiatives of individual officials at different levels. A more structured mechanism would promote optimisation of research applications.

On Extension support to the Societies field officers comprising of Area Officer and the Procurement Assistants would visit individual societies and provide technical guidance to the members on both crop technology and management of cooperatives. However, in actual practice, since the ratio of Procurement Assistants to Societies is in the range of 1:10 or more, the whole concept has been utterly diluted. The very large number of Societies spread over a vast area makes a mockery of extension support.

This problem is further compounded by the fact that the Procurement Assistant is also responsible for input support, implementation of Agriculture Departments' programmes under O. P. P., laying of demonstration, distribution of mini-kits, procurement of oilseeds, transportation of the oilseeds to the oil mills etc. Thus the very large area coverage as well as the multi-dimensional nature of the duties of the Procurement Assistants militates against a successful formal mechanism of regular extension support.

What came as a surprise was that inspite of this deviation between the originally commenced role of the Procurement Assistants in providing extension support and its role in actual practice, there seemed a reasonable level of client satisfaction. This was due to both an appreciation of the formidable nature of the task of the Procurement Assistant as well as a deep and intense disillusionment with the poor level of services rendered by the VEW of the Agricultural Department. The latter was virtually non-functional and the efforts of the Procurement Assistant stood out in stark contrast.

To sum up, some important conclusions that emerged from our study were as follows:

(1) GROFED has facilitated creation of village organisations in the form of Growers Cooperatives providing specialised integrated services to member farmers.

(2) Coverage of the Societies is thin as only 40 per cent of the farmers in the Societies' area have taken membership and only 9 lac ha. out of total 23 lac ha. area under oilseeds in the State is covered.

(3) No effective mechanisms exist for regular and formal research – extension linkage at various levels. This has been largely left to informal channels.

(4) Extension support was weak due to the fact that—

   (a) the procurement assistant was required to cover too vast an area and perform too large a number of functions, and

   (b) in many cases he was technicaly ill-equipped being a non-Agriculture graduate.

(5) The Societies provide potentially effective forum through which input management as well as extension support can be

effectively provided in case the number game becomes more favourable.

N.M. SADGURU WATER AND DEVELOPMENT FOUNDATION

Sadguru Water and Development Foundation is a Public Trust. The trust has two distinguished members on its Board of Trustees. An advisory council of 15 members comprising of Professionals, Administrators and Technocrats provides guidance in planning, organisation and execution of programmes. A full time Director heads the organization. He is assisted by Programme Executives, Programme Officers and Supervisers. The Foundation began its integrated rural development activities in 1976 for the benefit of the tribals of Panchmahal district of Gujarat State.

Objective

The overall objective was to improve the living conditions of the tribals. This objective was to be achieved through (a) development of land and water resources in the area, (b) development of environment, (c) Harvesting of rain water, (d) creation of additional employment in the area by increasing cropping intensity, (e) checking migration from villages.

Approach

The approach adopted by the foundation was consultative. Needs of Tribal farmers of the area were assessed through surveys, discussions and observations. An assessment of Land, Water, Human, Environment etc. was also done by Professional staff of the foundation and programmes were then developed by adopting farmer participatory approach.

Programme

The foundation is implementing lift irrigation, social forestry, agro forestry, watershed development, horticulture, women income generation programmes, bio-gas, preventive health services in Gujarat, Madhya Pradesh and Rajasthan.

Linkages

The foundation has strong linkages with the State Government Departments. From the beginning to 1990 funds for lift irrigation, other irrigation activities, social forestry etc. came from DRDA, Integrated Tribal Development Project. The Government of India agencies like CAPART and NWBB, the Madhya Pradesh Government also provided funds for lift irrigation. Private commercial companies like STANROSE provides funds as well as managerial support. Aga Khan Foundation gave grants for managerial support and for programme activities. NORAD is also supporting various programmes through Ford Foundation Grants field capability and training have been strengthened.

Foundation is also interacting with scientists working in Research stations of Agricultural Universities in Gujarat, Madhya Pradesh and Rajasthan.

RAJEDIA PROJECT

Twin villages of Rajedia and Kheda are located on the bank of river Machhan at a distance of 3 km of the Dhod-Banswada highway. It is a rolling country side inhabited by Bhil-Bhilala tribe. The tribal families are mainly dependent on agriculture for their livelihood. Prior to implementation of lift irrigation scheme the area was monocropped. Maize was the sole crop grown by tribals. Tribals were to migrate to cities in search of employment in post-monsoon season. Tribals are living on dispersed homesteads. A total of 155 families are living on their own holdings. Majority of the families own 1 to 1.5 acres of land. Farming, therefore, was primarily rain fed at very subsistence levels.

NMSWD foundation had undertaken a project of Lift Irrigation in Shankarpura which brought prosperity to the area. Residents of Rajedia who have been interacting with the residents of Shankarpura got motivated to see the results of Lift Irrigation Project which brought prosperity to fellow tribals of Shankarpura. The community approached Foundation for installation of lift irrigation system on river Machhan. The foundation's core staff visited the village and had in-depth discussions with the tribal farm families. The foundation staff first assessed the physical, social, environmental situations. The preparedness of the community to undertake and sustain the project was examined in detail. Technical surveys in regard to water availability, lifting of the water and its storage, distribution of the water to dispersed land holdings were studied. After assessing the technical feasibility of the project and formulation of the estimates, Department of Minor Irrigation, Government of Gujarat was approached to fund the project. In the mean time staff of the foundation continued to have discussions with the residents of the villages and worked out the modalities in respect of participation of farmers in the project. It was decided that all 155 beneficiaries will provide manual labour required for construction of a 3 meter high check dam, storage tank on the hill top, laying of the pipe line to fields and construction of water courses to the fields of farmers. The lift irrigation project completed in 1990-91. A cooperative society was formed of which all 155 farmers are members. Ten-member executive with a secretary and president is entrusted with the operation and maintenance of the lift irrigation system. Elections are held regularly. There is no voting but through consensus office bearers are chosen. The tribals belong to the same clan and therefore are a very homogeneous group. The President of the society Shri Rupa Bhai is 70 years old and commands lot of respect among the residents. He has been associated with the Foundation since 1973, when the Foundation Director Shri Jagawat was surveying the entire area on the banks of the river Machhan.

The lift irrigation enabled farmers to raise three crops in a year. Besides maize in monsoon, farmers cultivate wheat, gram in winter season and moong in summer. Availability of water, focussed attention of the farmers on problem of soil erosion, on their fields. The need to check soil erosion was discussed by the farmers with the foundation staff who assisted them in installation of lift irrigation system and also providing them with technologies of crop production under irrigated condition. The foundation core staff having expertise in soil water engineering after due survey prepared a plan of construction of field bunds, contour bunds, graded bunds and terracing of steep slopy lands. Farmers actively participated in the field surveys of their fields and in preparation of the plan.

The community members decided that the construction of bunds be done collectively on each farmers field on all 155 holdings and 25 per cent of the cost of construction of bunds and terracing, field channels etc. will be borne by them in the form of manual labour. Soil conservation work has been completed on all 155 holdings in both the villages.

Once upon a time both the banks of river Machhan was full of dense forest. Tribal people have great liking for trees. In past years there had been indiscriminate cutting of trees. With the availability of water the community decided to take up afforestation programme on the village waste land and also plantation of trees on their holdings. Foundation expert provided them with technical knowhow of raising trees of economic importance. A few of the farmers were trained in nursery raising with the assistance of forest department. Forest department also provided them with plants to undertake farm forestry and afforestation work on village waste land. Tribals preferred to plant eucalyptus, bamboo, accasia and other species which can be easily sold in the market. Similarly for farm forestry they preferred to plant eucalyptus as it can be sold in 2-3 years. Preference of fruit trees was for Ber, Guava and Mango.

The prosperity visible in Rajedia–Kheda has been attained by the tribal community through their own efforts under expert guidance. They have developed required will, capacity and ability to improve their quality of life and sustain the same.

Linkages

The foundation continues to have close working relationship with the tribal community even after completion of the three activities. The programme officer in charge of cooperatives and extension each month holds a meeting of the beneficiaries and discusses progress and problems and assists them in resolving the same. Information and knowledge about new technologies in agriculture, watershed management, forestry are passed on to farmers in these monthly meetings. The Secretary of the cooperative society is in constant dialogue with the Programme officer to sort out operational problems e.g., recently the motor of Lift Irrigation pump was burnt. The cost of rewinding of the motor was Rs. 10,000 the society had no money to pay for repairs. The Programme Officer arranged for loan from the foundation to be repaid after wheat harvest.

The foundation has an excellent working relationship with the State government. In fact it has high credibility in the government system. This helps the foundation to receive funds from State government for various projects.

At the operational level the linkages with the other government functionaries like Agriculture Department, and other Development Departments arc non-existant.

The following lessons can be drawn from this project:

1. If the project is need based the participation of beneficiaries is readily forthcoming. There is little difficulty in mobilizing the community.
2. Homogenity of the community accelerate the process of change as there are little chances of conflict which may hinder the process. If there are any, they quickly get resolved.
3. High professional competence of external agency is key to success of the project. In this case the foundation is supported by technically competent staff who are able to execute the project successfully.
4. Involvement of target group at every stage of the project promotes a sense of belongingness which in turn builds confidence in the community to own responsibility for implementation and sustenance of the project.
5. Emphasis needs to be given for maximum utilization of local resources – physical, biological, social, technical and human e.g., in this project local residents worked as masons, pump operators; stones available were used for masonry work.

KRIBHCO INDO-BRITISH RAINFED FARMING PROJECT

Krishak Bharti Cooperative Limited is a multi unit cooperative society. The production and distribution of agricultural inputs to farmers through the cooperative channel is the main activity of the Kribhco. It is set up as a cooperative enterprise. Its members are sales and purchase cooperative societies both at the national, State, district, tehsil and primary levels. The fertilizer is marketed through cooperative institutions and network of societies. Kribhco is also undertaking extension and research programmes. The promotional programmes were initiated in the year 1985-86. Kribhco field staff conducts, block demonstrations, farmers meetings, Field Days, Cooperative Conferences, Crop Seminars, Special Campaigns, Training Programmes, Agriculture Mechanization Programmes, Adoption of Cluster Villages, Opening Krishak Bharati Seva Kendra, Drought Relief Programmes and Development Projects in the marketing zone of the society in 12 States.

Under its development programmes, Kribhco has undertaken a new project with the assistance of British Government. The project is known as Kribhco-Indo-British Rainfed Farming Project—The Project is managed directly as a functional specific activity with a line reporting to Kribhco through a manager. The project is jointly funded by Overseas Development Administration (ODA) under its poverty focussed budget, along with Government of India through their respective State Governments.

Objective

The project aims to develop and implement a participatory and poverty focussed approach to farming systems development in a low rainfall, drought prone, environmentally degraded and largely tribal area of Western India. The specific objectives are as follows:

1. Develop and implement a participatory and poverty focussed approach to agricultural research and development;
2. Improve livelihoods of rural poor families;
3. Support the key role of poor and women in the integrated farming systems;
4. Establish village institutions in the Project Area which will sustain the process of participatory approach of agricultural development.

Organisation

The project is located in three districts of Panchmahal, Banswara and Jhabua in the States of Gujarat, Rajasthan and Madhya Pradesh. The project has selected six clusters to represent different socio-economic and agro ecological conditions, these are:

— Sarjumi-Jhamri Cluster (Limkheda North), Gujarat.
— Bor-Agaswami Cluster (Limkheda South), Gujarat.
— Jahrod Chatra Cluster (Kushalgarh West), Rajasthan.
— Bijori Cluster (Kushalgarh East), Rajasthan.
— Umaradara Cluster (Thandla East), Madhya Pradesh.
— Mandli Cluster (Thandla West), Madhya Pradesh.

Each cluster has about six to eight villages and in one year the project will work in four of them. High potential areas within the three districts are excluded from the project. The project headquarters is located in Dahod. All clusters are within a distance of 70 km from Dahod.

The project is managed by a special unit Nithin Kribhco. It has a team of specialists in both social and natural resource development who provide professional support to community organizers both men and women living and working in village clusters.

The project unit is headed by a Project Manager stationed at Dahod. Panchmahal district of Gujarat. The Project Manager directly reports to the Chief Manager (Marketing Services) in New Delhi.

Three social scientists besides their functional responsibilities are also coordinating the activities of the Project in the three States of Madhya Pradesh, Rajasthan, and Gujarat to enable to achieve unity of command in the project implementation. Whereas, technical specialists are functionally responsible for project activities in other functional areas, emphasis has been laid on creating an organisational structure which avoids hierarchical distinctions and is participative between field and office. Project also gives attention to the Human Resource Development and career needs of staff through periodic training and workshops. The staff has undergone three training programmes in PRA techniques, Technology especially Agro Forestry and sensitivity to situations.

Approach

The project aims to improve the livelihoods of poor families in the area by identifying, testing and making available agricultural technologies that are appropriate to poor farmers, especially scheduled tribes, scheduled castes and women. It will involve institutional development at village level and linking of participating households and communities to governmental and non-governmental organizations and programmes. It is envisaged that with the active involvement of local farming communities and drawing on jointly developed understanding of the traditional maize based farming system, the project will be able to prioritise problems to be solved and identifying opportunities for innovations. The focus is on meeting the needs of women and the poorest farmers through continued communities involvement implementation. The emphasis is on to generate a local capacity to manage resources in a better way and to obtain services and inputs from government institutions. The strategy adopted by the project is:

— Learning from experience of Village Communities
— Drawing upon farmers knowledge through close interaction

— Developing alternatives which have better adaptability and sustainability

— Developing capabilities within the communities to undertake such alternatives themselves. For this farmers are encouraged to play key roles in experimentation and judging the success of new technologies

— Create linkages with, on going projects, institutions in private, public sectors and NGO

— Involvement of women as equal partners in all project activities.

LIMKHEDA (NORTH) CLUSTER

The village Sarjumi-Jhamrian selected under the project in Limkheda (North) cluster have subsistence form of agriculture. The crops are grown under rainfed situation. The dominant crop is maize. In low lying fields paddy is grown. The uplands are utilized for cultivation of pigeon pea, groundnut. The area is basically monocropped. Gram and wheat are grown on medium soils with one or two irrigation from the nearby rivulet.

Cattle are the important livestock and castrated males are used as draught animals. Cows and Buffaloes are low milk yielders. Sheep, goat and poultry are also kept by a few farm families. People supplement their income by migrating to nearby cities in off season and also from collection of forest produce. Land holdings are small. Majority of farmers own less than 2 ha. of land. The forest cover adjoining the village cluster is diminishing. Non availability of fuel and fodder poses problem in their day to day living. Similarly, drinking water supply is inadequate as the village well is very deep with inadequate storage. Hand pumps installed by the Public Health Department remain out of order for want of the time.

The project community organizes one male and one female conducted a series of PRAs with the farm families. The settlement pattern of the families is in four localities. The PRAs were conducted in rotation in each of the four settlements when all farmers and farm women were invited to discuss various problems. Community organizers through personal visits to individual farm families established rapport with the community. Separate PRAs were also conducted for women to know their specific problems.

After each PRA, the project staff reviewed the process among themselves to ensure that the objectives of the PRAs are being attained and modified if required. Separate PRAs were conducted by women community organizers with the assistance of specialists and other co-workers. The PRAs first were conducted with small groups of women in all the four settlements. Problems were identified, prioritization was done and action plan was developed. For example drinking water supply was a major problem. The hand pumps were out of order, women were required to fetch water from long distance, although male farmers did not show much concern about this problem. Hence it was decided to get the hand pumps repaired on priority.

Meeting this felt need of women has accelerated the process of group action among woman. Special interest groups are being formed to pursue activities of their own interest. e.g., cattle management, plantation of fruit trees, budding of ber trees, health and education of children, vegetable cultivation in backyard etc.

PRA technique helped the community to identify the problems, fix priorities, develop plan of action, assign responsibilities on individuals to promote group action. For example residents of the village identified deforestation as a major problem. They collectively named 70 species of the trees in existence in the surrounding hills. They had shown their preference for 4 species for plantation on their lands and also on common village land. Similarly, in crop farming focus was on raising productivity of the rainfed maize having white seed and also of gram in rabi. The project staff organized visit of a few farmer representatives as decided in meetings to Banswara KVK, where they got the opportunity to see cultivation of new white seeded maize varieties and also of other crops. The discussions with the scientists at KVK motivated farmers to try new varieties of maize gram on their farm.

The project staff in collaboration with the scientists from Zonal stations of Banswara and Jhabua developed a programme of on-farm participatory research in the project area. A four stage approach to the identification of improved cultivar is being used.

1. Through participatory planning sessions with farmers. The main characteristics which farmers want in new cultivars will be assessed.
2. A search is being undertaken to identify a range of cultivars for all the main crops in the area which are acceptable to farmers.

In the research process the cultivars will be assessed in light of farmers perceptions, needs and priorities, also suitable to agro climatic conditions of the area. They would be selected from national releases, state releases, private sector releases, breeders advance lines.

3. Seeds of identified cultivars will be procured and multiplied.
4. Farmers will be invited to select from the cultivars offered and will jointly evaluate them with project staff.

Based on this on farm participatory approach number of cultivars of maize and gram are being tried on farmers fields. This process has generated interest among farmers in new crop production technologies.

Village women are also actively participating in developmental activities. Besides attending common meetings of male and female they are also participating in separate meetings being organized by the community organizers.

Thus the project staff has been able to initiate group process in the village. A stage has been set for formulation of a Gram Vikas Mandal through which all developmental activities will be implemented.

Lessons which can be drawn from this project

1. Adequate preparation is needed before undertaking a project, i.e., infrastructure, appropriate technical support, technically competent staff with aptitude to work in remote areas and adequate financial support.
2. Training mechanisms as per needs of the project area should be developed so that the staff develops sensitivity towards client situations, is able to use PRA techniques and technical and management skills.
3. Importance needs to be given to clients own experience and knowledge of his method of problem solving.
4. Informal, closer interaction among staff helps in building a congenial work atmosphere.

Linkages

The project has been able to establish good working relationship with Research, Government Departments and NGOs, LMP, KVKs in Rajasthan are providing support for training in agriculture technologies, forestry, soil conservation to field staff and also farmers.

Scientists are participating in on-farm research.

Sadguru foundation has agreed to assist in installation of Lift Irrigation Project.

AKRSP
(AGHA KHAN RURAL SUPPORT PROGRAMME)

A Case Study of Farmers' Participatory Approach

The AKRSP is a non-communal non-profit organisation set up in 1983 to serve as a catalyst and stimulant for rural development in India, with special emphasis on income generation and better management of resources. A donation received from Agha Khan forms the corpus fund. It receives periodic donations from the Agha Khan Foundation in India and avails substantial assistance from relevant State Government schemes.

The basic objective of the AKRSP is alleviation of rural poverty, with special concern for the weaker sections of rural society.

This, AKRSP seeks to achieve by organising rural people into viable and self-sustaining institutions. It is their belief that it is these grass-root level organisations that enable people to build up a capacity to determine their own priorities for development, take joint action to implement these priorities and gain access to resources and services available from Government and other agencies.

The parameters that have been laid down for programme implementation are:

(1) It would not function as a funding organisation but would provide direct support to village institutions.
(2) It would promote income generating activities and not welfare programmes such as those related to health and education
(3) Development activities would be routed through the village institutions. AKRSP would not extend its organisation to the village level but provide support to appropriate village institutions which would decide, execute and manage the development programmes.

The operational area of AKRSP is initially focussed on selected backward districts of Gujarat where it is working in about 150 villages. The programme in which AKRSP has been providing support to village organisations are:

(a) Water Resources Development viz., Community Lift Irrigation Schemes, Percolation Tanks and Check Dams.
(b) Watershed Development and Management including Watershed Development and Soil and Water Conservation in private lands.
(c) Agriculture both irrigated and rainfed
(d) Savings, credit and supply schemes
(e) Animal Husbandry
(f) Marketing
(g) Biogas

The AKRSP has Dr. I.G. Patel as Chairman and an eminent Board with Mr. Anil Shah as Managing Director. It has a team of professionals in the areas of Forestry, Irrigation, Engineering, Agriculture and General Management at the Central Office in Ahmedabad. The three programme areas of Bharuch, Surendranagar and Junagadh each have a Spearhead Team comprising Programme Coordinators, Programme Organisers who are professionally in various fields numbering about 55. Envisaged as a support structure, the Spearhead Team is in constant touch with the programme villages and, in turn, receives support from the core team in programme formulation and implementation.

In all AKRSP supported projects, there is emphasis on five essential features:

(a) peoples' participation and training
(b) technical excellence
(c) cost effectiveness
(d) equity: special concern for the weaker sections, including women
(e) replicability

AKRSP's approach is to commence with a series of dialogues by its staff with the villagers. They are informed about the objectives of AKRSP, the types of programmes it supports and the manner in which the support can be availed by them.

Prior to the actual selection of the project, villagers are assisted to consider various alternatives and development opportunities and choose those which appear the most feasible and desirable to them. To facilitate this process, AKRSP conducts Rapid Rural Appraisal to assess the state of local resources, peoples' perspective of their use and their priority for developing them. Thereafter, the villagers are required to form a village level institution called the Gram Vikas Mandal (GVM), membership of which is open to all.

Village representatives are taken to visit completed AKRSP works so as to see for themselves the nature of activities and the manner of support provided by AKRSP. On deciding the schemes the villagers want to take up, an agreement is entered into between GVM and AKRSP which clearly lays down the responsibility of the village institution and AKRSP.

Village Institutions usually take the responsibility for mobilising local human resources for the execution of the project. They select from amongst themselves, village "extension volunteers" who take training relevant to the projects selected and when completed, manage them. AKRSP usually takes the supporting responsibilities of preparing plans and estimates that would be acceptable to the concerned Government agencies, mobilising financial resources from Government and other sources. It also trains the local village volunteers to take up responsibility in programme implementation, provide financial and administrative support during the gestation period and ensure proper management of the projects when they are completed.

We had visited two villages in Bharuch programme area of AKRSP, the first was Samarpada hamlet in Bilothi village which had a homogeneous composition of 55 tribal households all of whom were members of the GVM which was constituted around 1987-88. The pattern before AKRSP entered the area was that since all lands were mortgaged to one big farmer, all villagers would migrate to urban areas in search of employment for a good part of the year. However, with the introduction of a land redemption scheme and a bullock credit scheme, supported by AKRSP, villagers ceased to migrate and, instead, started working on their own lands which now provided them gainful employment.

This being an older village, P.R.A. techniques were not used in a formal sense, yet, however, through consultations with the GVM, programme prioritization was made and social and farm forestry works, soil and water conservation works and rainfed agriculture works were taken up. Works were explained to the villagers in the GVM, volunteers identified and responsibilities assigned to them. Though initially very few members of the GVM came forward to take up works on their lands, in course of time, with the gentle persuasion by the volunteers as well as the visual impact of the works on other farmers' fields, a much greater response was generated.

With the increase in number of works the Extension Volunteer started developing a greater stake as he began to derive income ranging from Rs. 2000 to Rs. 6000 per year paid jointly by the individual beneficiary and the community through the GVM. Some volunteers were then designated as Master Extension Volunteers on the basis of the higher level of skills displayed by them and the higher output generated.

In Ghodi village covering 95 farmers over 315 acres Lift Irrigation Scheme costing Rs. 12 lacs had been commissioned in 1989 with AKRSP support.

The scheme which distinguishes itself for its technical excellence is also a role model of farmers' participation. The Society which manages the scheme meets once every year to elect an Executive Committee, which, in turn, elects a President. The day to day affairs of the Society are handled by a paid Secretary who draws Rs. 500 from the Society's account. The Society also pays Rs. 270 for an Assistant Secretary, Rs. 450 each for a Pump Operator and Water Distributor. The entire cost of operations and management is recovered from the member farmers on a differential rate from season to season on per acre per metering basis. Interest of 10 per cent is charged on delayed payments. There are hardly any defaulters and the Society is desirous of building a cash reserve of Rs. 1 to 1.5 lacs for meeting contingencies after which they could as well do without AKRSP intervention. All the personnel have been provided training through AKRSP and appeared highly motivated. The whole village after switching over from a single rainfed crop to three irrigated crops was now a sea of prosperity and boasts of 60 cycles, one tractor, foodgrain self-sufficiency, higher consumption of vegetables, 8 steel almirahs, and better percentage of schoolgoing children.

Some of the important conclusions that appeared from our study are as follows:

(1) AKRSP has done a remarkable job in trying to develop village institutions be they GVM in rainfed areas or Lift Irrigation Societies or Mahila Mandals. These Village Institutions, which have been expanding their operations from agriculture, watershed development, forestry etc., to commercial ventures

like input distribution, marketing are harbingers of hope for the future.

(2) Though Lift Irrigation Schemes by virtue of the very strong economic interests of the members are likely to be indisputably sustainable, there seem some reservations about the sustainability of the GVM in the rainfed areas, as and when AKRSP withdraws support or programme funds are not forthcoming from Government or any other agency.

(3) The panacea for this seems to be to strengthen the external linkages of the GVM's with rural banks, DRDA's, Forest Departments, suppliers of seeds and fertilisers – a process which AKRSP staff is currently engaged in.

(4) The concept of "Extension Volunteers" evolved by AKRSP seems to have struck a right chord in the developmental process. This institution, if properly nurtured by systematic selection of EV's regular training, their relevance to the community, consequent cost-sharing of their services between the individual beneficiary farmers and the village Institution, is likely to increase the schemes sustainability.

(5) The insistence of AKRSP in getting the farmers to pay, however nominally, whether in cash or kind, for all services rendered to them has increased their sense of responsibility and several instances of recoveries above 80 per cent can be found in loan based schemes at around the same point of time when Government was operating its loan waiver scheme.

INSTITUTIONAL ARRANGEMENTS IN AGRICULTURAL EXTENSION—
A COMPARATIVE ANALYSIS

		KVK		NGO	NGO-GO Collaboration		Private Sector	Professional Association	Co-operatives
Sl. No.	Salient Features	Sri Aurobindo (NGO)	ICAR	Myrada (Kamalapura)	Myrada DLDB (Kamalapura)	ASW	ITC-ABD (Sunflower)	Agri-Horticultural Society	Fresh Marketing Co-operative Society
1	2	3	4	5	6	7	8	9	10
1	Objectives	1. Training – Farm Family – Farm Youth – School drop-outs – Farm Women – Master Trainer-cum Entrepreneur	1. Training – Farmer – Farm Youth – School drop-outs Farm Women 2. Farm Advisory	1. Experimentation with sustainable development process 2. Institution building process	1. Develop 'Software' for participatory integrated watershed development	1. Facilitate collaborative programme among NGOs and GOs. 2. Experiment with farmer Participatory methodologies to:	1. Integrated Oilseeds development 2. Value-added processing 3. To produce quality seeds	Promotion for agriculture and horticulture.	Extension services to vegetable and fruit growers 2. Procurement of farmers vegetables and fruit produce

(Contd.)

1	2	3	4	5	6	7	8	9	10
		2. On-farm Research for Integrated Farming Systems 3. Livelihood systems 4. Farm Advisory		3. Participatory resource development and management		– Control RHC – Seed village	4. To promote sunflower cultivation		3. Supply of vegetables to consumers at fair price
2.	Programme thrust	1. Crop Husbandry 2. Horticulture 3. Sericulture 4. Income generating activity	1. Rainfed agriculture 2. Horticulture 3. Income generating activity 4. Machinery Maintenance	1. Participatory resource appraisal, planning and development 2. Appropriate local level institutions development	1. Building people's institution 2. Influencing DLDB's mode of hardware development	1. Blending of informal and formal R&D 2. Self revolving seed village	1. Hybrid seed development 2. Popularisation of sunflower technology 3. Market intervention operation in sunflower seeds	1. Promote gardens of economic and decorative value 2. Supply and services points 3. Organisation of garden shows	1. Establishment of 4 procurement centres 2. Establishment of 33 retail sale 3. Promotion of hilly vegetable cultivation

(Contd.)

1	2	3	4	5	6	7	8	9	10
		5. Livestock crop integration		3. Rural credit systems			4. Integrated oilseeds processing	4. Conduct professional lectures	
		6. Machinery Maintenance		4. Training				5. Promote professional fellowship	
								6. Promote processing storage and marketing activities	
3(a)	Who designs	Designed by professional (founder)	Researcher designed	Professional and farmer in collegiate mode	Professional and farmer in consultativein mode	Professional and farmer collaborative mode	Professional designed contractual	Professional designed designed	Cooperative personnel
3 (b)	Who implements?	Implementation by trainee with backup support of NGO.	No proper follow-up	Farmer implemented	DLDB implemented	Collaborative implementation	Farmer implementation under close professional supervision	Professional implemented	Representative Management Committee implemented

(Contd.)

1	2	3	4	5	6	7	8	9	10
3	(c) Who manages?	Trainee (farmer) managed	–	Farmer managed	DLDB managed	Farmer managed	Professional led but farmer managed	Professional managed	Representative Management Committee managed
3	(d) Monitoring and Evaluation	Performance linked evaluation	–	Farmer and professional	Farmer and NGO partner	Professional with farmer	Professional	N.A.	Centralized
3	(e) Feedback	Trainee's feedback responded	–	Farmers's feedback incorporated in improving next phase of planning and implementation	DLDB insensitive to feedback	Farmers feedback used to improve experimental design	Farmers feedback used to launch aggressive promotion of Hybrid seeds	Responsive to growers' feedback on supply and services	Moderately sensitive to feedback
4	Outcomes	1. Improved livelihood 2. Skill upgradation	–	1. Empowerment 2. Resource regeneration	1. Empowerment 2. Capacity Building	1. Farmers capacity to experiment enhanced	1. Increased income	1. Educational	1. Popularization of vegetable cultivation among tribals and around cities

(Contd.)

1	2	3	4	5	6	7	8	9	10
		3. Enhanced productivity		3 Capacity building 4. Asset creation	3. Erratic implemen-tation	2. Integrated pest Mana-gement 3. Availability of quality seed 4. Farmer-to-farmer diffusion	2. Greater awareness 3. Market development		and towns. 2. Regular vegetable supply to city.
5.	Research Extension Linkage Mechanism	Informal	Strong Link-age	1. Training research and exten-sion staff 2. Informal research linkage	1. Poor Linkage	1. Strong linkage with research and extension	Utilitarian linkages with research and extension.	N.A.	Linkages with other marketing societies of other states.
6.	Gender-Sensitivity	High	Low	High	Moderate	Low	Low	Low	Low

ITINERARY OF CONSULTANTS

Date	Institution	Place	Activity	Resource Persons
1	2	3	4	5
08.02.93	Agri-Horticultural Society	Hyderabad	Visit to Nursery	Dr. Anantha Reddy Additional Director, Horticulture
	Fresh Marketing Cooperative Society	Hyderabad	Marketing of Vegeta-bles and Fruits	Sh. M. Balakristaiah Managing Director Fresh
09.02.93	ASW	Waddicherla	Control of R H Cater-pillar, Seed village programme	Mr. M.A. Qayum Mr. M.V. Sastri Mr. Vittal Rajan
- do -	ASW	Madipadiga	Untreated village No control mea-sures against RHC taken up	
10.02.93	ASW	Wanaparthy Komalpally	Untreated village	
11.02.93	Sri Aurobindo Institute of Rural Development	Gaddipally	KVK	Dr. G. Gopala Reddy Dr. R. Reddy
12.02.93	CRIDA-KVK ICAR	Hyderabad	KVK-CRIDA	Dr. Subba Reddy
13.02.93	Myrada	Kamalapur	Myrada Project	Mr. J. Jangal and Staff
14.02.93			Ninghe Nala Watershed, Wadgera Mini Watershed, Margutti Sub-watershed in collabora-tion with DLDB.	
15.02.93	ITC-ABD	Hyderabad (F.N.) Kurnool	Discussion with Dr. Ranga Rao-ITC.	

(Contd.)

1	2	3	4	5
16.02.93	ITC-ABD		Journey and Discussion with Shri Vithal	
		Sri Ram Nagar	Sunflower Demonstration	
		Nagireddy Palli	Sunflower Hybrid seed Production	
		Nandyalarmpet Duvvur	Sunflower Demonstration plot	
		Allagadda	-do-	
17.02.93		Kurnool	Visit to seed quality control, Lab, Seed Processing Unit	
			ITC Research Farm Salkapuram	
			ITC Agro-Tech Mantralayam oil extraction and Refinery Unit	
		E. Tandrepadu	Visit to ITC Research Farm.	

ANNEXURE VII

GUIDELINES ON FARMER PARTICIPATORY STUDY

FIELD VISIT TO DIFFERENT AGRICULTURAL SITUATIONS

K.C. John, C.S. Rajan, D.K. Sharma and *Charanjit Singh*

Setting

In India a greater part of agricultural development takes place within the context of State-led agricultural extension services. National Extension Services (NES) block that coincided with Community Development era encompassed both agricultural and non-agricultural as well as extension and non-extension activities. Subsequently, centrally designed projects like: Intensive Area Development Programme (IADP), Intensive Area Agricultural Programme (IAAP), High Yielding Varietal (HYV) Programme and more recently Special Crops Production Programme and Technology Missions represent a continuing trend of experimentation to evolve an effective vehicle for transferring agricultural technology in the form of information and subsidised inputs. However, most far reaching and comprehensive reforms of extension are embodied in the Training and Visit (T&V) System.

Reforms in Extension Services

T & V in its ideal formulation, conceived full time commitment of the village extension worker (VEW) to dissemination and advisory tasks, requiring him to be relieved of all other responsibilities. It also instituted a single line of command within each State Extension Service. The T&V System also emphasizes links with other agencies and programmes. Especially, with the advent of National Agricultural Research Project (NARP) and establishment of Zonal Research Stations (ZRS) on agro-climatic zone basis, the linkage between extension-research was not only decentralized but also strengthened substantially. It also sought to institutionalize mutually beneficial tie in involving regular working meetings of SMS's and scientists of State Agricultural Universities (SAU's) and joint diagnostic studies. Besides, the logic of T&V's operations also necessitated establishment of two-way linkages between extension and suppliers of credit, irrigation water, seeds, fertilizer, power, farm machinery, pesticides and marketing services. Regular informal contact with administrative and revenue officials and with key officials and functionaries of the line departments – soil conservation, animal husbandry, horticulture, forestry etc. – were envisaged. By thus structuring and animating organic relationships between agencies, T & V System was to lay the foundation for an innovative approach to agricultural development.

T & V System sought to tackle a number of chronic problems associated with agricultural extension services – poor organization, dilution of effort, poor coverage and mobility and limited training—with an approach that stressed a unified service, concentration on extension, regular farmer training and frequent agent-farmer contact.

T & V, however shared a common perception with other traditional extension services that the stimulus for change in farming communities must originate from outside and that this change takes the form of externally generated knowledge and technologies that will modernise traditional ways of cultivation. With its emphasis upon externally generated technologies, hierarchical bureaucratic structure and rigid framework of training and farm visits and its obsession with quantitative results, the T & V system in reality is conspicuous for its lack of meaningful farmer participation and its inability to respond to different agricultural situations. The frantic way in which the T & V System has been implemented in different States has left little scope for processes such as participation. Apart from a few exceptions, by and large, the farmer is viewed as the mechanism by which objectives and targets can be achieved. In this scenario, farmer participation and situation-responsiveness, is likely to be perceived as an extra burden on an already over-worked village extension worker.

Need for Innovations in Extension Services

Extension professionals, however, have begun to argue that the T & V System must not be a one-way process and that the village extension workers must be willing to learn from and work with the farmer in a participatory manner. The rigidity and the enforced external discipline of the system has been critically commented and there has been a general call for a loosening and for allowing a more genuine two-way flow of ideas and decision making within the system.

The major issue for the extension services in terms of farmer-participation and situation-responsiveness is how to turn around decades of established practices. The FAO's *Agricultural Extension Manual* suggests that the extension workers' task of promoting farmer participation is not an easy one but offers the following recommendations on how it might be achieved:

- Farmers are more willing to participate in activities which meet their felt needs. A quick needs assessment can determine farmers' needs and priorities. Again, the needs of all people should be taken into consideration, not just those who are accessible and cooperative.
- If farmers are encouraged to express their needs and provide some input into the structure of a project/programme, they should not be ignored. The price of widespread participation may be an additional burden on the extension worker, but if she or he wishes to sustain farmers' involvement, then the farmers' ideas must be taken into account.
- Farmers are more likely to participate if actual benefits are directly tied to participation.
- Farmers, especially those with low incomes, are more likely to participate and remain involved if the benefits are material, direct and immediate. There is evidence that in working with the uneducated, benefits have to be obvious and tangible. One of the best ways of getting farmers' interest is through the use of convincing and realistic demonstrations and trials (Swanson 1984, p. 67).

Rhetoric and Reality

Despite the efforts of such authors as Jiggins and Roling (1982), Belloncle (1987) and Roling (1987) to fundamentally reexamine extension practice, it could be argued that the above is characteristic of current extension practice in terms of participation. In this practice, participation is essentially functional and its implementation a mechanical exercise. Ironically, despite its solid grounding in non-formal education and its emphasis upon farmer as an adult who has views and ideas, current extension practice reflects anti-participatory approaches. The rhetoric of extension is diametrically opposed by the practice which basically is still unable to treat the farmer as an equal. Jiggins and Roling (1982) while confirming this rhetoric also examined how emerging new approaches to extension, from a technical change to an institution-building approach, were beginning to come to grips with the issue of participation. They further suggested that a more participatory approach to extension must include such elements as : user influence and control, group approaches to extension clientele, mobilisation and organisation of small farmers and new styles of collegial management.

Lessons from Farmer Participatory Projects

Our visits to farmer participatory projects during February 8-17, 1993, have sensitised us to the following four main principles of the methodology of participation:

- Emphasis on the process of participation rather than immediate quantitative outcomes. Various initiatives promoting participation have shown flexibility and willingness to experiment and did not allow the pressures of immediate, quantifiable impact to undermine the process of participation.
- Ensuring a balance between awareness creation and economic activities since both are important and indeed mutually supporting. Neither awareness with no tangible objective nor short-term tangible gain with no substantial base for longer-term participation are sustainable. Overemphasizing economic activities on the assumption that these will naturally provide the base for future participation is a sure recipe for failure.
- Institution building and community mobilisation to ensure a secure and locally available foundation for future activities, as well as minimising the inevitable external dependency.
- Maintaining regular contact between the people and professional staff, since participation is a labour-intensive process and develops better where there is continuity. External catalytic support is fundamental to the process but it must be both reliable and regular. A process of participation can only take a hold if in the early stages, there is this regular contact;

otherwise it becomes limited to periodic responses or contributions.

It seems that the emerging methodology of participation is different from conventional project methodology used in public extension services in a number of important ways; it is less constrained by time, it is less concerned with immediate quantitative impact, it is less ready to tell the farmer what to do and it is less committed to a predetermined line of action. It also demands qualities, skills, resources, time and imagination that are not in abundance in the public sector extension systems. From successful farmer-participatory projects, it was also evident that a methodology of participation is not a one-off affair, but a continuous unfolding of practice until a state of genuine participation is achieved. Their success also depended on the preparations that they make to develop the local community capability and the institutional reforms within the agency to become more participatory and the investment in experimenting and developing an appropriate methodology.

The issue, therefore, is how conventional top-down extension practices can be changed to encourage the wider influence of these emerging methodologies. Can methodologies being developed in the NGO sector be transplanted into the public system? Realistically it might prove an impossible task. Alternatively, one can scout for individuals within the public extension/research services who have intuitively embraced farmer-participatory processes and disentangle the variables – policy, organizational structure, tasks and processes – that might aid in institutionalizing farmer-participatory processes in the extension services.

This would necessarily mean operational understanding of participation within the extension services, if any, and will demand alternative methods of information and data collection and analysis.

Indicators of Participation

The widespread support for the notion of participation is yet to be translated into developing operational indicators or guides to how this participation might be assessed.

Successful individual or system initiative in farmer participation in the public extension/research services would be identified on the following indicators of participation:

Qualitative Indicators

1. Clients' role in problem diagnosis and planning phases:

    (a) degree of participation in the problem formulation phase
    (b) degree of participation in generating ideas
    (c) degree of participation in planning
    (d) client commitment to project

2. Clients' role in implementation phase:

    (a) degree of financial contribution
    (b) degree of indigenous knowledge vs. dependency on outside experts.
    (c) degree of participation in implementation
    (d) degree of organization of clients
    (e) extent to which organization is their vs. engineered by others
    (f) democracy and equality in organization
    (g) extent to which clients redesign project

3. Clients' role in maintenance phase:

    (a) degree of participation in maintenance
    (b) degree of indigenous knowledge used vs. dependency on outside experts
    (c) degree of local vs. outside ownership and control of facilities and organizations

4. Systems linkage to clients:

    (a) adequacy of communication from extension worker
    (b) degree to which clients' capacity has been increased

Perceptions of farmer-participatory professional regarding organizational factors that either impede or facilitate participatory process would be elicited.

Stakeholders Perceptions

Besides understanding innovative professionals approach to farmer-participation during the field work, conscious effort would be made to discern important stake-holders' perceptions on feasibility of experimenting with farmer-participatory research/extension methodologies. Stakeholders will consist of farming community from representative villages of the selected agricultural situations, extension and research groups in the region and key officials in the State Directorates of Agriculture.

Participatory rural appraisal techniques will be employed to discern farmers' analysis of their agricultural situation, factors that reduce productivity of agricultural/livestock enterprises, inefficient resource use, inappropriate enterprise-mix and farmers' practices that are not sustainable in the long run. Farmers' views on extension/ research systems will also be elicited through Venn diagrams and access, control, benefit matrix.

In Workshop environment, the consultants will seek views of extension and research professionals on important issues impacting on sustainability of local production systems and juxtapose them with farmers' perceptions. Suggestions on making extension/research systems farmer-participatory will also be sought.

Middle-level professionals at District level and senior level officials DOA will provide insights into policy issues, structural aspects and linkage mechanisms that might facilitate experimentation on farmer-driven and client-accountable approaches.

BROAD ISSUES ON WHICH RESPONSES WILL BE GENERATED FROM RESPONDENT GROUPS BY EMPLOYING P.R.A. TECHNIQUES OR PROBING ETC. IN THE COURSE OF ONE MONTH FIELD VISITS TO VARIOUS STATES

Farmers

1. What are his needs in the sphere of Agriculture including Horti-culture, Watershed Development and Soil Conservation, Farm Forestry, Animal Husbandry etc.?
2. How does he fulfil his needs at present?
3. What is his understanding of the present system of agricultural extension and research? How does it operate? What is its coverage both in terms of number of farmers and programmes?
4. To what extent does he feel that the present system of

agricultural extension meets these needs, either partially or substantially? And how?

5. What in his mind are the ways in which the present agricultural extension and research systems can be made to cater to his needs more effectively?
6. Does he favour an approach involving greater farmers' involvement in the operation of these systems? If so, what are the possible mechanisms through which such approaches can be formulated? Whether through any existing village institutions (VI's) or by creating any new VI's?

Extension Workers

1. What are his perceptions of the needs of the farmers in the areas of Agriculture including Horticulture, Watershed Development and Soil Conservation, Farm Forestry and Animal Husbandry etc.?
   (Juxtapose farmers' perceptions here to compare the perceptual difference in needs)
2. To what extent, in his mind, is the present system of agricultural extension and research fulfilling these needs?
3. If the system is not substantially addressing these needs?
4. What are the Extension Workers' needs, in terms of

   (a) Training – Is he satisfied with the present mechanisms? If not, what improvement does he have in mind?
   (b) Information, Knowledge and Technology – What are the sources from which he draws them?
   (c) Administrative-Infrastructural facilities, methods and procedures etc.

5. How effective are the present systems of agricultural extension and research linkages?
6. If not adequately effective, what are the reasons?
7. How can they be made more effective?
8. What are their views on evolving a model of agricultural extension and research which is more farmer participatory?
9. How can such a model be evolved? What are the mechanisms through which it can be institutionalised?

10. What role does the Extension Worker see for himself in such a set up?

Researcher

1. What are his perceptions of the needs of the farming community in the sphere of agriculture including Horticulture, Farm forestry, Watershed Development and Soil Conservation, Animal Husbandry?
   (Juxtapose the perceptions of the farmers and extension workers and compare the perceptual differences)
2. What are his perceptions of the extension functionaries needs from the Research System?
   (Juxtapose the perceptions of the farmers and extension workers and compare the perceptual differences)
3. To what extent is the Research System taking care or meeting these needs of the farmers and the extension workers?
4. If not adequately, then why not? What are the reasons?
5. How can it be made to more effectively meet these needs?
6. How effective are the present systems of research extension linkages?
7. If not adequately effective, what are the reasons?
8. How can these linkages be made more effective?
9. What are his perceptions of the structure of the ICAR system and its linkages with the University System?
10. Does he favour evolving a model of agricultural research which is more farmer participatory? If so, in what areas of research?
11. How can such a model be evolved? What are the mechanisms through which it can be institutionalised?
12. What role does the Extension Worker see for himself in such a set up?

ANNEXURE VIII

PROCEEDINGS OF THE SECOND NATIONAL WORKSHOP OF THE STEERING COMMITTEE CONSTITUTED BY THE GOVERNMENT OF INDIA HELD ON APRIL 2 AND 3, 1993 IN MANAGE, HYDERABAD

Government of India has constituted a Steering Committee to guide the consultancy on making recommendations for restructuring the existing agriculture, research and extension systems in the country with a view to make them more farmer participatory. The first meeting of the high level Steering Committee was held on 1st-2nd February 1993. In the second meeting of the committee the Consultants, upon return from their field visits, presented a draft set of recommendations which was subject to critical evaluation by the committee.

Since the discussions centred around a few broad issues, a presentation of the discussion is being made issue by issue. Copy of the list of the participants is also enclosed.

Issues for Discussion in the National Workshop in MANAGE on 2nd and 3rd April 1993

1. How to operationalize the concept of participation
   - Technology Generation
   - Technology dissemination
   - Empowering farmers groups
2. Alternate roles for extension
   - Skill requirement
   - Other institutional support
   - Gender differentiated extension
3. Guiding principles of group formation
4. Accountability issues Extension Research

- Rewards and incentives
- Responsiveness

5. Appropriate management system
6. How to operationalize cost sharing principles
7. Role of NGO's

Alternate Roles for Extension

Participants in the national workshop held on April 2 and 3, 1993 at MANAGE, Hyderabad endorsed the proposition that research, extens·n and farmer be considered as sub-systems of an agriculture information system rather than as separate systems in themselves. The need for two way communication between the three sub-systems was also recognised in contrast to the present situation where researchers have little contact with farmers. In effect research and extension processes have to be more farmer participatory.

T & V system of extension is focussed on crop related technology transfer by VEW to identified contact farmers through a schedule of visits. The T & V approach is being uniformly implemented throughout the country without any differentiation of farming situations. Research and Extension needs of farmers practicing agriculture in rainfed, irrigated and commercial areas are quite different as not only are there variations in the nature of agricultural enterprises but also in the socio-economic levels of farmers. Further, in rainfed farming system areas, the need is to make extension programmes more broad based rather than keep the focus on narrow crop production technology to generate employment and income levels of farmers.

Similarly the emphasis would have to be given to the human resource development to improve technical managerial skills of all categories of farmers especially the neglected ones i.e. women, youth and landless labour. This can be achieved effectively through the process of group mobilization and institutional building to sustain the process of change.

Extension is being called upon to perform a different role in irrigated and commercial areas as the needs of farmers in these areas are quite different. Demands for new economically viable technologies and wide array of second generation problems have arisen as a result of indiscriminate use of inputs. These needs cannot be met by passing on routine messages during VEWs visits as he is inadequately equipped for the task. Specialised information and guidance is needed to cater to the needs of these categories of farmers to adopt team approach for problem solving.

In effect extension system will have to perform the following alternate roles in different agricultural situations:

Rainfed Farming Systems

1. Transfer of technology through suitable extension packages, not only related to crop production but of all land based enterprises.
2. Institutional building at the village level through the process of group mobilisation and act as catalyst between groups.
3. Human resource development to promote technical and managerial skills among all categories of farmers. The focus has to be on less resourced, under privileged groups like women, youth and landless labour.
4. Provide feedback and assist research system for development of location specific technology through diagnosis of problems, problem formulation and on-farm research
5. To assist research system to become more farmer participatory.
6. To establish linkages between line departments for transfer of technologies for land based enterprises.
7. Assist line functionaries of agriculture department in services and supplies. It is recognised that in remote areas where market infrastructure is poorly developed, services and supplies will continue to be channelled through extension VEWs. However, emphasis should be on creating a balance between extension and non-extension functions as both are two sides of the same coin and important from the stand point of raising productivity and production. Similarly, it is also recognised that grant of subsidies to poor resourced categories of farmers is to be viewed as an incentive to adoption of technology. Transfer of technology is to be fully supported by availability of inputs to bring about adoption of technology

Irrigated Areas

1. Assist research system to identify problems and needs
2. Assist in monitoring of sustainability issues.

3. Assist in creation of data base trials
4. Dissemination of information
5. Mobilisation of group for area based problem solving

Commercial Areas

1. Advisory and Consultancy
2. Promotional

Skill Requirement

In order to perform these roles, extension personnel need to possess the following skills:

1. Diagnostic skill
2. Skills in group processes
3. Skills in human resource development
4. Technical skills
5. Skill in information processing

Institutional Support

Appropriate institutional support is crucial for successful performance of envisaged roles of extension personnel which may be listed as follows:

1. The research system has to provide technological support.
2. Line functionaries in agriculture department have to provide required administrative and financial support.
3. Line departments related to broad based extension programmes have to provide technical manpower and related technology.
4. Input and credit agencies in private and cooperative sectors have to provide active support by arranging timely supplies and services.
5. NGO's support is needed in institutional building, methodology dissemination, articulation of farmers demand on research services, and training of farmers and extension workers in group dynamics.

Gender Differentiated Extension

Need for focus of research and extension systems to women farmers is being increasingly realised as they have special technological and educational needs.

The number of women headed households are increasing due to seasonal and permanent migration of men to non-farm employment in urban areas. Most of the women operate at subsistence or below subsistence levels. Absence of adequate productive resources and institutional support results in low levels of productivity. Apart from women headed households, the other rural women are actively engaged in agricultural production systems. They need access to extension and related services to increase their efficiency and reduce the drudgery of agricultural operations that are classified as "women's work".

The extension task, therefore, be focussed on the following aspects:

1. Group mobilisation
2. Involvement of women in different stages of appropriate technology development
3. Focus on transfer of appropriate technology
4. Information processing as per their need
5. Assisting concerned agencies to organise supporting services and supplies

Guiding Principles of Group Formation

With the present system of extension it is difficult to meet millions of farmers individually. Therefore, it is essential that these farmers are organised into small groups at grass-root level for receiving the benefits of development.

It is increasingly realised that provision of economic inputs alone will not solve their problems. Many of the problems of resource poor farmers are the result of oppressive social structure. Mobilizing these farmers will empower and energize them to take effective decision and actions which will lead to improved production and productivity. Small farmers are eager to adopt new technologies, however, they are often handicapped by their lack of self image, courage and resources. Many small and marginal farmers lack confidence in the decisions they make, often they look towards others for confirming their decisions. Under these circumstances farmer groups will provide support and encouragement to take effective decisions. The group approach to extension work and services will remove the drawbacks and bottlenecks of traditional approach of extension system which is highly individual oriented. Organising farmers into groups will facilitate the effective functioning of extension organizations and other input institutions in different ways:

1. Various services can be made available to farmers through a single point, thus avoiding the duplication of services and confusion arising out of multiplicity of agencies.
2. Farmers in groups can be effectively assisted, with farm advisory services, marketing facilities and transport arrangements.
3. Extension personnel and groups can actively interact and identify technological problems and needs of farmers.

Farmers group can select members to go for special training on different subjects, on their return others insist that all the new information and practices are immediately passed on to them. With this method adoption is rapid and widespread, providing the hypothesis that farmers learn most rapidly from other farmers.

Group extension methods provide better feedback which makes it possible to reduce some of the misunderstandings that may develop between the extension agent and farmer. There is also greater interaction between the farmers themselves. Group discussions also play an important role in increasing knowledge and changing attributes as well as in changing behaviour.

Farmers groups can be organised in different ways:

- Resource rich
- Resource poor
- Similar socio-economic conditions
- Land holding basis

  (a) Small farmers
  (b) Marginal farmers

- Youth groups
- Women farmers
- Various interest groups

- Groups on basis of contiguous fields
- Landless
- Watershed basis
- Enterprise based groups
- Groups on basis of irrigation supply by canal or tube well, horse power basis
- On economic motives
- On basis of technological innovations
- On basis of specific technology requirement
- Homogeneous voluntary groups out of existing panchayats
- Farmers who would like to diversify the cropping pattern to get higher income

While forming the group it may be kept in mind that it has a good feedback mechanism to find out the gaps in their technological needs. As the feedback required is of different kind, various types of groups could be organized to have specific feedback. Before the group formation process is initiated there is need to identify activities where problems cannot be individually solved. Benefits should be more than the costs involved. Effective group formation is a time consuming process.

The following initiatives are to be taken to start the work of organising farmers in small groups:

1. Policy decision to contact farmers in groups instead of individual contact be taken and should be strongly supported adminis-tratively.
2. Groups should be demand driven and not project driven.
3. To start with work of group formation be taken up where some NGO is working.
4. This work could also be started where some society like milk producing society or poultry farming society is functioning.
5. Disseminating the message to neighbouring villages about the advantages of group formation.
6. Size of the group be kept small and in no case it should be bigger than what the members want.
7. Demonstrate the process of group formation by organizing visits of groups of farmers to already developed active farmers groups in nearby areas.

There are no specific guidelines for the formation of farmers groups. The community should be stimulated to form groups by explaining to them the advantages. There should not be rigidity whether the group is formal or informal. The staff involved for farmers group formation should be carefully selected. They should be committed workers and should have exposure to behavioural sciences. Such workers could be sent for two to three months training. Regularity, discipline and professionalism in managing the groups has to be ensured.

Cost-Sharing

While appreciating the broad thrust of the recommendations, the participants felt that, in the present Indian context introduction of the principle of cost-sharing in the public agricultural extension systems was perhaps premature given the limited economic capabilities of the majority of Indian farmers as well as the high level of dissatisfaction with the quality of extension services available at present. It was felt that an essential pre-requisite for cost-sharing was the emergence or evolution of a more credible extension system and hence the emphasis should be more on improving the quality of public services thereby improving the prospects of wider acceptability of the principles of cost-sharing.

Role of NGO's

The recommendations relating to the proposed involvement of NGO's in agricultural extension were unanimously endorsed by the participants. It was recognized that the roles of Government and the NGO's would largely be mutually complementary as one would strengthen the weaknesses of the other. In particular, NGO's could be mobilized in a big way in developing farmers organizations at the village level, given their superior organizational and social capacity building skills. However, a note of caution was struck by some participants who referred to the uneven quality and distribution of NGO's as also the possible resistance in being "drafted" into collaboration with Government unless they perceived that Government had something credible to offer. Amongst the subjects in which it was felt that NGO's could train extension personnel, Natural Resource Management deserves inclusion in the light of the invaluable experience gained by

NGO's. Also they could play an important role in articulating farmers' needs with a view to influence public policy. Another role suggested was that of assigning 'Action Research' or 'Operations Research' project to NGO's.

Appropriate Management Systems to Increase Accountability

Institutional changes envisioned in making public agricultural research and extension systems farmer-participatory and client-accountable should be translated into appropriate management systems. Three major areas of extension/research management that would require immediate attention are: decision making, communication and management control. Issues discussed included:

(a) Centralized decision-making should give way to decentralized programme planning at the local level,
(b) Increased capacity of extension to provide proper feedback and reduce message distortions between research and farmers community,
(c) Appropriate kinds of communication support systems are required to enable the agricultural research and extension systems to be farmer participatory,
(d) Supportive role of electronic media and indigenous communication methods to reduce extension costs,
(e) Increased role of farmers and farmer groups in agenda setting, and formulation of standards and objectives which serve as a guide for performance,
(f) Involvement of clients in monitoring and evaluation of performance of extensionist according to standards set,
(g) Institution of appropriate incentives and rewards system to reinforce farmer participatory behaviour among extensionist and researcher, and
(h) Designing appropriate institutional mechanisms to enable mid-term corrections that are contingent on findings of client-participatory monitoring and evaluation.

Research – Extension Linkages

The existing structure of research-extension linkages comprises:

(a) A national level Research and Extension Coordination Committee headed by Director General, ICAR with representation from ICAR, Ministry of Agriculture, Planning Commission, NGO's etc. for the purpose of coordinating research and extension functions at the central level and developing policy guidelines in this regard.

(b) A State level Research-Extension Coordination Committee chaired by Agricultural Production Commissioner and comprising the Vice-Chancellor of SAU's, Secretaries and Directors of allied departments, representatives of NGO's, farmers and others to review functioning of research-extension coordination mechanisms at the State level and consider issues arising out of ZREAC for formulating policy guidelines.

(c) A Zonal Research and Extension Action Committee for the purpose of operationalising research-extension linkages. This Committee is established at the Z.R.S. level in each agro-climatic zone and is chaired by Director Research/Extension of the SAU. It consists of the Additional Director (Extn.), Department of Agriculture, Zonal Joint Directors of the Agricultural Departments, Zonal Coordinator, Directors, State Departments or their nominees, Heads of KVKs/FTCs heads of sub-stations. This committee is to meet pre-seasonally to identify technological and socio-economic constraints in transference of location specific technologies and develop an action plan for the same, to review feedback from extension and the solutions offered to the same by the researcher, to plan the schedule of monthly workshops and visits of the joint diagnostic teams for the forthcoming season and to practise and review the conduct of farm trials for the zone.

(d) Monthly workshops wherein the Master Trainers from the Z.R.S. train the SMS of the Agricultural Departments. This mechanism serves as the clearing house for resolving problems and issues faced by extension and research systems and also plans the extension programme for the coming month.

(e) Joint diagnostic visits by teams of scientists from the Z.R.S. and Extension Officers from the State department once during each season.

LIST OF PARTICIPANTS FOR THE SECCND NATIONAL WORKSHOP HELD ON APRIL 2 AND 3, 1993

Dr. N.K. Sanghi
Zonal Coordinator
CRIDA, HYDERABAD

Dr. J.C. Kalla
Director
NAARM
HYDERABAD

Dr. Amar Singh
Joint Commissioner (Extn.)
Ministry of Agriculture
NEW DELHI

Mr. Pravin Srivastava
Joint Commissioner
Ministry of Agriculture
NEW DELHI

Dr. V. Ranga Rao
Executive Vice President
ITC, HYDERABAD

Dr. S.B. Ghosh
Director,
NARP, ICAR
NEW DELHI

Dr. John Farrington
Coordinator
Research and Extension Network
Agricultural Administration Unit
ODI, LONDON

Dr. K.V. Raman
Ex-Director, NAARM
HYDERABAD

Dr. A. Appa Rao
Ex-Vice Chancellor, APAU
HYDERABAD

Mr. James Mascarenhas
OUTREACH
Myola Villa
No. 28/2, Bangalore

Dr. Mane
Programme Executive, Agril.
AKRSP, AHMEDABAD

Dr. Kamal Kar
Training Organiser
SEVA BHARATI, KAPGARI
MIDNAPORE DIST.

Dr. Kalkat
No. 706, Sector 11-B
CHANDIGARH - 160011

Dr. Y.P. Singh
Professor (Agricultural Extn.)
IARI, NEW DELHI

Dr. J.S. Kanwar
Ex-DDG, ICRISAT
HYDERABAD

ANNEXURE IX

DRAFT RECOMMENDATIONS

FARMER PARTICIPATORY STUDY*

1.0 Background

Government of India commissioned a Farmer Participatory Study to review the existing structure of agricultural extension and research and suggest measures to improve quality, management and cost effectiveness. More specifically the study team received the following brief:

1.1 To generate alternative options that encourage farmer participation in planning and management of agricultural extension and research.

1.2 To suggest institutional mechanisms to make research scientists and extension professionals accountable to the client community; and

1.3 To design pilot projects for empirically evaluating feasibility of different options of farmer-driven and farmer-accountable agricultural extension and research.

2.0 Methodology

2.1 The study team adopted an interactive and iterative method of dialogue with a specially constituted steering committee. In the first instance, a set of critical issues germane to the study were identified and a two-day brainstorm held with the expert panel. Suggestions emanating from the brainstorm session were appropriately incorporated in the study plan.

---

* This brief note was prepared to stimulate focussed discussion with the Steering Committee during the National Workshop at MANAGE on April 2-3, 1993

2.2 The study team embraced a learning mode while visiting farmer participatory projects in Gujarat, Karnataka and Andhra Pradesh. Essential ingredients of farmer participatory projects in the co-operative, agro-input industries and NGO-sector were distilled as well as their replicability in the public sector extension and research examined. The lessons emerging from the farmer participatory projects are:

2.2.1 *Participation as a process:* Successful farmer participatory projects emphasised the process of participation as opposed to immediate quantitative outcomes. They also demonstrated flexibility and willingness to experiment.

2.2.2 *Balance between human resources development and transfer of technology:* Instead of adopting a narrow view of extension, farmer participatory projects ensured a balance between human resources development and transfer of technology since both aspects of extension practice are important and indeed mutually supporting. Neither awareness building and education with no tangible objectives nor short-term tangible gain through transfer of technology with no substantial base for longer-term participation are satisfactory recipes.

2.2.3 *Institution building is a pre-requisite for farmer participation:* Farmer participatory projects were built upon a local organisation to ensure a secured and locally available foundation for future activities as well as minimising the external dependency. Institution building activities included mobilisation, organisation, training, technical support and system management.

2.2.4 *A team approach:* A common feature of many participatory projects is the presence of a project team which ensures genuine client participation. External catalytic support is fundamental to the process; but reliability and regularity are critical. Given the complexity of the process of participation and the fact that in its initial stages at least it is labour intensive, a team approach has advantages over the lone field worker. A team allows both for the sharing of experiences and for the mobilisation of resources.

2.2.5 *Stake building:* Projects that promote participation have invariably designed experimentation with stake building exercises in the local organisation. It may take the form of membership fee, proportionate contribution from the wage income and usufruct rights, savings and thrift etc. Besides building a corpus for the local organisation, it encourages regular contact and discussions among local community as to better ways of using funds and thereby provides continuity in situation analysis, local responses and stakeholders contribution. This process also enables owning up of the local organisation and promotes emergence of functional leadership.

2.3 The second phase of field visits involved interactions with farming communities, researchers and extension professionals of distinct agricultural situations. The agricultural situations visited were:

(a) assured irrigated areas (command areas in Tamil Nadu, tarai region of Uttar Pradesh, tube-well irrigated plains of Western Uttar Pradesh and shallow tube-well irrigated alluvial plains of West Bengal),
(b) assured rainfall region (Vidarbha region of Maharashtra,. Brahmaputra valley in Assam and Coastal areas of West Bengal and Tamilnadu),
(c) hilly regions (Meghalaya, Kumaon and Garhwal ranges in Uttar Pradesh and the Nilgiri Hills),
(d) rainfed region (drought-prone lateritic regions of West Bengal and Tamilnadu, and
(e) arid zone (desert of Rajasthan).

Participatory Rural Appraisal (PRA) techniques were employed to access farmers analysis of their environment and current and emerging agricultural problems; obtain farmers perceptions on efficacy of existing institutional support and feasibility of their meaningful participation in agricultural extension and research.

Interactive workshops with groups of researchers and extension professionals helped the study team to understand the functioning of existing structures and their strengths and weaknesses. Several options of making the existing extension and research systems farmer participatory and farmer-driven were also explored.

2.4 The present workshop provides an opportunity to the study team to use the expert steering committee as a sounding board and test the implementability of the emergent sets of recommendations on a pilot scale in distinct agro-ecological situations. The draft recommendations would be suitably revised to incorporate critical suggestions received from the august steering group.

2.5 The final report will be critically reviewed by relevant officials from Government of India, State Governments, State Agricultural Universities and Indian Council of Agricultural Research. In a national workshop key officials would come together to comment on the salient recommendations of the study team and suggest plausible follow up actions.

3.0 Diagnosis

3.1 The study team, based on its analysis of agricultural situations and differential research and extension needs, has classified the country into following generic groups (Table 1):

TABLE 1: **Classification of Agricultural Situations in India**

Distinctive Features	Assured Irrigation System	Rainfed Farming System	High Value Commercial Crop System
1	2	3	4
1. Literacy	High	Low	High
2. Economic Status	Affluent	Poor	Affluent
3. Production System	Simple Monocropped	Complex Diverse Enterprise	Specialised Commercial Enterprise
4. Dominant Occupation	Crop Husbandry	Crop Hus-bandry and livestock/ Horticulture/ Aquaculture	Crop/Tree Husbandry
5. Risk Factor	Low	High Drought/ Flood Prone	Low to Medium
6. Ecological Fragility	Low	High	Low to Medium
7. Topography	Usually Flat Lands	Usually undulating lands	Usually Flat Occasi-onally Hilly
8. Soil Status	High Fertility	Problem Soils	Fertile
Special Problems	Waterlogging Salinity	Acidic Soils	
Erodability	Low	High	Low to Medium

(Contd.)

TABLE 1: *(Contd.)*

	1	2	3	4
9.	Input Consumption	High	Low	High
10.	Productivity	High mean and Low variation	Low mean and High Variation	High mean and Low variation
11.	Role of Indigenous knowledge System	Obsolete	High	Low
12.	Social conditions	(a) High evolved farmers pressure group (b) Highly articulate individual farmers	(a) Low capacity among farmers to articulate requirements (b) Weak farmers organizations, NGOs, help to speak on behalf of the poor and for the long term strengthen (a) & (b).	Highly evolved farmers' organizations; Highly articulate individual farmers
13.	Illustrative list of States/Regions	Punjab Haryana Western U.P. Krishna-Godavari belt of A.P. Thanjavur belt of Tamil Nadu Central Bihar	Rajasthan Gujarat M.P. Eastern U.P. Bihar Bengal Karnataka	Himachal Pradesh Parts of Punjab Parts of Maharashtra Parts of A.P. Parts of Tamil Nadu, Kerala

1. Assured irrigated areas
2. Rainfed areas
    (a) Rainfed crop-based farming system,
    (b) Pastoral systems,
    (c) Rainfed horticulture-based farming systems,
    (d) Rainfed aquaculture-based farming systems, and
3. High value commercial agriculture regions.

The study team is of the view that given the well developed structure of agricultural extension and decentralized research in India, issues like appropriate role of extension in different environments, the financing of services and the links between extension, research and farmer knowledge should be addressed immediately. Furthermore, use of contact farmer, purist transfer of technology, impact messages, and extension impact warrant reexamining from a wider context in which extension operates and móve towards some of the specific conditions under which extension is likely to prove effective and capable of being sustained financially.

3.2 Situation Analysis

3.2.1 *Assured Irrigated Areas*

Two of the major sources of agricultural growth in the most favoured areas – the spread of modern varieties and rapid increase in fertilizer consumption – have already been substantially exploited. Genetic gains in yield potential in successive generations of improved varieties have slowed and increasing proportion of research resource is now being devoted to "maintenance research" to protect yield gains against breakdown of resistance to pests and diseases.

Though major sources of growth, viz., genetic gains, fertilizer-responsiveness and assured moisture regime have been exploited to a large extent, substantial opportunities exist to increase production through increased yields, reduce costs through efficient use of purchased inputs and water and selection of efficient and improved cropping systems.

A wide array of second generation problems have begun to emerge in these areas that impact on sustainability of production systems. Increased yields and cropping intensity coupled with farmers practices of imbalanced nutrient management has tended to deplete soil of both macro and micro-nutrient reserves, i.e., nutrient mining.

Cereal: Cereal rotation and control of broad leaf weeds has reduced the competition for grassy weeds. As a result incidence of specific weeds have increased. The changing weed-regime and their increasing population demands an integrated understanding and weed management strategies. The continued practice of cereal—cereal or vegetable—vegetable rotation also creates conducive environment for pest pathogen build up in the soil. Indiscriminate use of chemicals, in turn, has induced emergence of newer bio-types and strains that are resistant to recommended chemicals.

Increased supplies of tubewell water, flat-rate of water pricing and higher cropping intensities complicate water and salinity control management in irrigated areas. Seasonal and permanent draw down of acquifers in different irrigated areas raises serious questions about sustaining the level of productivity and production in the grain-baskets of India.

Farmers have not been able to keep abreast of these changes in the technical environment. Productivity in the most favoured region has declined overtime due to depletion of soil nutrients, pathogen build-up draw down and weed-regime. Research on sustainable agriculture should be accorded high priority.

The second generation inputs often require more information and skills for successful adoption than the initial round of introduction of miracle seeds which were well known for their wider adaptability and nitrogenous fertilizer. The use of more complex fertilizers, micro-nutrients and pesticides often interacts strongly with variation in soil type and with year-to-year variation in micro climate and pest incidence. Hence individual farmers need to adapt the technology to their own requirements. Moreover, all second generation problems demonstrate common pool resources characteristics requiring collective action. Mobilizing of collective action, however, will depend on heightened awareness among farmers about their interdependence. In the absence of National Policies to ensure efficient input use, rather than designing external enforcement mechanisms; inter-dependent individuals, if facilitated, can institute self-binding norms and are capable of self governance of local resource management institutions.

Farmers' technical skills, information processing ability and management complexity required are of higher order than the first round of diffusion of innovation strategy. The value of traditional knowledge systems and intimate knowledge of their environment in the science-based agriculture tends to depreciate rapidly as new problems emerge and subsequently new inputs and cropping patterns are introduced.

The transfer of technology (TOT) mode of experiment station research used by agricultural scientists and the technocratic progressive contact farmer approach used by extension service appear inadequate models to guide us in addressing the sustainability issues.

3.2.2 Rainfed Areas

While there is likely to be a stronger demand for qualitatively improved technical and farm management advisory services in the assured irrigated area the four identified rainfed systems pose major agricultural challenges in the design and conduct of research in agricultural systems which are poorly-understood and, to a large extent, resistant to conventional extension approaches based on a series of commodity and component recommendations. Farming community rely on agriculture practised under, what has now come to be characterized as the complex, diverse and risk-prone (CDR) conditions for their livelihood. For instance, in rainfed crop-based farming systems area the farming community has to cope with moisture stress, marginal problem soils and undulating topography with high run-off in the drought prone regions; in the flood-prone basins the problems are abundance of water due to drainage congestion and waterlogging and unpredictable rainfall that may induce drought and flood conditions in a single production cycle.

On the other hand, pastoral systems are primarily based in marginal desert areas of Rajasthan and Gujarat. The pastoral community experience frequent food deficits, were among the hardest hit during the successive drought years in the 1980s and their production systems suffer serious short, as well as long-term perturbations. Consequently the community is regularly on the move in search of greener pastures. On-going environmental degradation and fragility of the ecosystem calls for a re-orientation of extension toward environmental protection and enhanced resource conservation activities with the potential of augmenting the local livelihood strategies.

Specificities of the hilly horticulture-based farming system arise due to interactions among features like inaccessibility fragility, marginality, heterogeneity, micro-niches and human adaptation mechanisms. Their implications for research and extension remain to be understood in a holistic sense. Sectoral extension programmes of soil conservation, reforestation, vegetable and tree-crop production and agricultural production have short-term focus, ad hoc basis, are resource extractive in nature and replicative of designs and experiences of plains that are often untested and unsuited to the hilly terrain. Both for pastoral and hilly horticultural systems a holistic natural resource management perspective rather than crop-based production systems is likely to yield rich dividends.

Coastal areas (West Bengal and Orissa) with recurring incidence of incursion of saline water pose yet another set of challenges to resource-poor farmers in eking out livelihood. Small fragmented holdings, high population density, high pressure to raise paddy, preponderance of rice pests and diseases, water logging, unsuitable water irrigation and saline affected patches compel the farmer to rely on mono-cropping with traditional pisciculture. While high-productivity potential aquacultural technologies have been generated under controlled conditions of experiment stations they have not been integrated to farming situations prevalent locally.

Increasingly, it is becoming obvious that mainstream agricultural research and extension strategies associated with the Green Revolution are not appropriate to more difficult and risk-prone rainfed areas. The unfavourable marginal and complex biophysical environments in rainfed areas, multiple objectives of farmers that is embedded in their survival strategies and complex interactions among the enterprise portfolio cannot be simulated on the experiment station. Nor will the current obsession with discrete component research with yield as sole evaluation criteria assist farmers in improving their existing production systems. Further, they operate under a complex socio-economic regime with subdued and interlocked markets that are highly unreliable. As a result, extension strategies that promote package of technologies not well matched to the inherent complexities of local farming systems are likely to be rejected.

Of late, a consensus seems to have been reached that livelihoods under rainfed regions can be enhanced through approaches which are systems-oriented and participatory. The usefulness of farmers knowledge, the partnership between farming community and research/ extension professionals, emphasis on identifying farmers problems as a starting point for participatory technology development, a broader rather than a "model" approach to research and extension are some of the salient characteristics of the emerging approaches to research and extension.

3.2.3 *High Value Commercial Agriculture*

Akin to the industrialized agriculture a few pockets of prosperous zones have emerged in India where farmers production systems are geared to cater to emerging markets—domestic as well as export. Floriculture, high value fruits—grapes, apples, organized orchards, organized aquaculture units, sugarcane and cotton are some of the illustrative examples.

Technical skills, information processing ability, managerial competence, market responsiveness and entrepreneurial skills among such farmers are highly developed. This group has the capacity to demand requisite extension services and the ability and willingness to pay for such services.

4.0 Recommendations

4.1 *Reconceptualising Extension and Research as Integral Components of Agricultural Information System*

Instead of viewing extension in isolation, it is useful to conceive the relationship among user (farmer), extension and research as sub-systems of an agricultural information system (Roling, 1988) with three major two-way communication channels between them. The objective of the agricultural information system should be to improve the decision making of the user sub-system. Extension cannot be improved without also improving research and mechanisms linking it to extension. Similarly extension cannot be improved without improving user control of farmers over extension.

4.2 *Transformed Role of Agricultural Extension*

The T & V approach of extension is narrowly focussed on crop-related technology transfer. But coping with the emergent problems of various agricultural situations call for major skill upgradation of farmers. Therefore, the goals of public agricultural extension systems should be enlarged to include both technology transfer and human resource development dimensions within its agricultural extension policy. While technology transfer contributes to increased agricultural production and enhanced national food security, human resource development increases not only the technical and managerial skills of farm households but also increases their ability to cope with risks and uncertainties and provide off farm entrepreneurial opportunities. These roles are not mutually exclusive.

4.3 *Farmer Involvement in Agricultural Extension and Research Policy*

Agricultural Extension and Research Policy should be consistent with and supportive of the recently announced national agricultural policy. Farmer involvement through formal mechanisms and informal consultations in policy formulation and periodic review is an effective means of creating "demand driven" national agricultural extension/ research system. It should be responsive to all major groups of farm people and flexible to encourage concerned public, private and nongovernmental organisations to contribute to the agricultural development goals.

4.4 *Gender Differentiated Clientele of Agricultural Extension System*

To broad-base agricultural development in the country will mean reallocating extension resources and developing a range of educational programmes that can transfer appropriate technology and develop technical, managerial and human development skills to all types of farmers on a more equal basis. However, subsistence level women farmers, inclusive of female-headed farm households and young farm families have special technological and educational needs. These two disadvantaged clientele require extension programmes that are specially targetted to their needs.

The number of women-headed households are increasing, specially in CDR regions, due to seasonal and permanent migration of men to non-farm employments, particularly in urban areas. Most of these women operate at or below the subsistence level and are also responsible for both the productive and reproductive needs of the farm household. They have competing demands on their time, low access to productive resources and institutional support, which results in low levels of productivity. They need extension programmes that will increase their productivity and access to resources that will increase their income levels and quality of life. Apart from women-headed households, the other rural women who are actively engaged in agricultural production systems need access to extension and related services to increase their efficiency and reduce the drudgery associated with many agricultural practices that are classified as "Women's Work"

Rural youth – out-of-school and educated unemployed – at present receive disproportionately inadequate share of extension services. Most of them lack basic agricultural skills. This next generation of farm families should receive higher priority from extension systems. Ramakrishna Mission's (Narendrapur) innovative effort in utilising rural youth clubs and training their members to become extension volunteers/rural entrepreneurs not only deserves praise but is also worth ådapting to different situations.

4.5 *Proactive Role of Clientele*

No one extension approach is appropriate for all conditions, objectives or clientele. However, the basic principles of successful extension system may be common. Extension Programme priorities should be based on both an assessment of client needs and client participation in programme formulation. The client groups as well as key collaborating agencies, should be involved in reviewing and providing feedback on extension programme content and approach as a means of improving performance.

4.6 *Importance of Farmer Groups in Client-driven Extension/ Research Systems*

User groups including farmers' organisations, farmer clubs, commodity growers association, young farmer clubs, special interest groups and the like are effective institutional devices in the creation of a client-driven agricultural research and extension system. Extension can play an important role as a catalyst in organising these farmer groups. These groups then can become important mechanisms in articulating specific research and extension needs, in facilitating horizontal diffusion, in accelerating technology dissemination and in developing technical and managerial skills and knowledge of farmers who participate in these organisations.

4.7 *Strengthening the Technical Content*

The technical content of agricultural extension programmes should be tested and validated to reflect the specific agroeconomic conditions of major target groups to be served in each agroecological region. T & V methodology provides for specific channels of feedback and communication between farmers and researchers. However, in practice, extension staff pass undifferentiated recommendations to farmers making no adjustments to their varied economic capacities and different farming systems. Recommendations are often prompted not by farmers expressed needs but by the procedural and academic imperatives that drive much of the experiment station research even at the Zonal

Research Stations.

Extension subject matter specialists are often insufficiently expert to modify these recommendations to fit local requirements. This pattern of advice is a symptom of break-down in the extension process. The study team felt that the critical break in the user-extension-research agricultural information system occurs not between research and extension, but rather between the research and extension sub-systems on the one hand and user groups on the other.

In the next section, we suggest methodological protocols that would break down unnecessary institutional barriers, identify specific problems and involve farmers in designing solutions to them. This also provides for in-the-field collaboration between researchers and extensionists on technology generation. Such on-farm training can eventually turn extension's subject matter specialists into true experts. Since, one of the inherent strengths of T & V is its ability to organize Village Extension Officers (VEOs) for the broad dissemination of information, it can effectively transmit the relevant results derived locally from participatory research.

4.8 *Farmer Participatory Approaches*

The following operational steps will enable farmer participation more meaningfully:

(i) Mapping of agricultural situations and inventorying of production problems. Participatory Rural Appraisal (PRA), techniques will help extensionists become acquainted with range intensities, context, problems and opportunities. This step helps in knowing what we don't know but they (farmers) know. (Involvement : Farmers/User Groups, Extension and NGOs).

(ii) Identification of homogeneous agricultural situations (Involvement: Extension, Research, NGOs)

(iii) Farmer Situation Analysis and Problem Formulation, Representative sites from homogeneous agricultural situations are selected. Employing PRA tools, constraints and opportunities of different types of rural households are analysed. (Involve-ment: Farmer, Extension, Research, NGOs).

(iv) Problem Prioritization: The first three steps form the diagnostic phase. Its outcome is a better appreciation of farming situations and identification of a large number of problems.

Participatory problem prioritization based on the criteria of regional annual productivity losses helps in ranking problems, inflicting largest losses. Effective problem solving will ensure greatest incremental contribution to research/extension efforts. (Involvement: Farmer, Extension, Research, NGOs)

(v) Planning and Designing Technology: This step involves linking formal and informal R & D to develop possible solutions to farmers' problems, given their needs, opportunities and institutional environment (Involvement : Farmer, Extension, Research, NGOs)

(vi) On-Farm Research Implementation: Depending on nature of prioritized problems, studies can be initiated either at the experiment station or in the farmers' field. While experimental station research is researcher designed and researcher implemented the recommendations will await findings of on-farm research and adaptive trials.

Four generic types of farmers participation in implementing on-farm research have been identified (Biggs, 1989).

Contractual	Scientists/Extensionists contract with farmers to provide land or services. It is Researcher-designed and Researcher-Implemented trials on farmers field.
Consultative	Scientists/Extensionists consult farmers about their problem and then develop solutions. It is Researcher-designed and Research-implemented trials. Farmers inputs are solicited at the diagnostic phase.
Collaborative	Scientists and farmers collaborate as partners in the research process. It is Research/Farmer-designed and Farmer-implemented trials.
Collegial	Scientists/Extensionists strive to strengthen farmers' informal research and development system in rural areas. It could take two forms : encouraging Farmer-designed and Farmer-implemented trials for prioritized problems or, Farmer-designed and Researcher-implemented trials to unravel the underlying logic of the indigenous knowledge systems.

(vii) *Multi-location Field Trials*: The on-farm research for technology generation is restricted to a few number. Based on its results larger scale multi-location trials are designed for fine tuning of technology. Fine-tuned technologies, in turn, can be laid out as demonstrations. Multi-location trials and demonstrations should be implemented in collaborative mode.

(viii) *Evaluation and Feedback*: At different stages of on-farm research, multi-location trials and demonstrations, farmers – both collaborators and non-participants – are provided with opportunities to evaluate the experimentation employing their own criteria. Researchers not only elicit farmers evaluation criteria but also incorporate the same in their subsequent experimental strategies.

The participatory approach also provides farmers a mechanism of expression of their needs to research, extension and policy makers through the diagnostic phase and of their evaluation of new technologies. The protocol thus encourages an effective two-way communication channel.

The feedback, thus generated is used for reformulating research, research agenda and formulation of extension messages.

(ix) *Dissemination*: Technology recommendations emanating from the farmer participatory approaches are transformed into different types of extension messages. Some of them are designed to inform farmers of the new recommendations, others aim to motivate farmers to apply the recommendations. Still others instruct farmers on how to put into practice the recommendations.

Monitoring farmer-to-farmer diffusion pattern and their adaptation behaviour will also assist in formulation of messages that transforms recommendations into farmer behaviours.

4.9 *Promoting Farm Extension Volunteers*

Two contentious issues have been raised often by critics of T & V System. First concerns the use and selection of contact farmers in T & V. Contact farmers are usually selected by the extension personnel and are supposed to be representative of socio-economic groups that exist in a particular local society. However, in practice, the study team found that invariably local elite pressures and staff preferences bias selection in favour of the richer and more influential farmer. The extension hierarchy in different States felt that such contact farmers not only do not transform extension messages into actual practice but also that they are least interested in transmitting technology to follow-up farmers. In such a situation farmer participation cannot be fostered.

A second contentious issue is the cost of T & V. The T&V approach to extension relies primarily on visits by VEOs to individual contact farmers. It envisages a major build-up of personnel in the extension services both at local level – for implementation – and at regional level – for supervision and monitoring. The concentration of effort associated with successful T&V therefore tends to be expensive. At several discussions, the issue of sustaining the budgetary burden, once the external donor assistance is removed, was raised. Further it was brought to the team's notice that a large number of positions remained vacant due to ban on recruitment.

Suggestions to increase the efficiency and effectiveness of extension veered round the concepts of local farmer organization and farm extension volunteers. Importance of farmer group has been dealt earlier.

The local organisation identifies among its members a few potential innovators and entrepreneurs for receiving training to upgrade technical skills and managerial ability. Such trained community members besides experimenting and practising improved farming systems, also assume the role of farm extension volunteers. This participatory strategy can overcome the problem of vacant positions by having larger number of extension volunteers whose background is appropriate to local needs. This together with purposeful participation and group pressure will increase the relevance of extension messages to local needs, enhance learning by user groups; provide relevant feed forward and feedback information and local analysis to ensure appropriate recommendations from agricultural researchers and extension staff.

Farm extension volunteers are accountable to the local organisation. Focussed training on appropriate technical skills, technology and communication methods is provided by extension staff. Suitable collaboration from NGOs can be solicited. Resources required in form of a token honorarium to put the concept of extension volunteers into practice can be provided to local farmers organisation on a tapering basis. Additional help in mobilising loans for setting up service facility for the volunteers i.e. input distribution, custom hiring etc. is envisaged. It is expected that eventually the local users would be able to bear the cost of volunteers' services.

Instead of acting as a "delivery system" the public extension system, henceforth, can help to organise local people to develop their own "acquisition systems", to acquire from the external environment those technologies and services that they want. Through this process, 'control' of extension through programme formulation and extension volunteers is passed on to the ultimate clientele.

4.10 *Accountability and Responsiveness*

Accountability and responsiveness are two sides of the same coin. Accountability addresses the issues of the effectiveness of staff, as well as effectiveness of programme. Responsiveness, on the other hand, denotes ability of system to respond to the real needs of farming community and respond fast enough to the dynamic socio-technical environment surrounding local farming systems.

In the agricultural information system, user groups can hold the extension sub-system accountable directly and the research sub-system indirectly through the farmer organization.

The participatory approach coupled with farmers' organization and extension volunteers ensure that the programme planning is controlled locally. VEOs agenda is formulated locally. Participation by representatives of extension, research and NGOs will contribute significantly to its success and legitimacy. But the decentralization of programme planning control allows priorities to vary as per the local needs.

The VEO is provided backstop technical and financial support to implement locally decided programme. Responsibility is matched with decentralized decision-making authority. Extension volunteers supplement VEOs. In villages with no VEO, the extension structure assists the farm extension volunteers to implement locally designed extension programmes.

In addition to the hierarchical authority, farmers' organization is involved in monitoring and evaluation of local extension staff. Thus, the current monitoring indicators, such as "number of visits made to farms", the "number of meetings arranged", and the "number of demonstrations organized" may change to "impact per person", "impact per unit area", or "effectiveness in solving some of the prioritized problems.

Research programme planning at the Zonal Research Station

(ZRS) is to be indirectly held accountable to the farming community. At present, research at the ZRS is sponsored either through the All India Co-ordinated Programmes, ad hoc or Agricultural Products Cess Fund projects of Indian Council of Agricultural Research, or through State research funds. The ZRS researchers should advocate modification of respective on-going co-ordinated programmes in the light of regional prioritized problems and be encouraged to prepare fresh research proposals of problem-solving nature for ad hoc and A.P. Cess fund grants from the ICAR. The State research resources should be allocated exclusively for adaptive research on farmers' field on a contractual basis. The prioritized problems at the district and region, consolidated by the DOA in close collaboration with ZRS researchers forms the basis for research resource allocation at the State level.

4.11 *Incentives and Reward Systems*

Though the Monitoring and Evaluation (M & E) unit regularly monitors the performance of VEO, most States are conspicuous by the absence of proper incentives and reward systems to reinforce desirable performance. Nor have many States given thought to career advancement prospects of VEOs. The study team observed the morale of local level extension staff waning.

Decentralization of decision making and flexibility in programming should be reinforced through instituting appropriate performance linked incentives and reward systems for VEOs.

Due weightage should be given to VEOs performance evaluation at the local organization level by the Department of Agriculture in designing a graded system of rewards. The graded reward system envisions reinforcing VEWs farmer participatory behaviour and the impact thereof, by citations of merit, a monetary incentive and awards based on annual performance review.

4.12 *Improving Communication Support*

Radio and electronic media can be effective and efficient means of disseminating certain types of technical and marketing information and of creating farmer awareness of new technology. Efficiency and effectiveness of field staff in group meetings and discussion, adaptive research/demonstrations and farmer field days can be greatly enhanced through proper use of teaching aids, audio-visual and other mass media material and extension publications.

A clearing house mechanism is also required at regional level for processing, indexing, abstracting and preparation of reader friendly documents on innovative extension strategies, approaches and methods and extension management for extension staff at different levels in the public, private and non-government sector. Developing and supporting such extension communication unit at the regional levels should be viewed as an essential component of an effective farmer participatory agricultural extension system.

4.13 *Educational Qualifications and Training Needs*

The study team found that the educational qualifications of extension personnel, for historical reasons, have remained very low in relation to the assignments and responsibilities they are expected to carry out. The monthly and fortnightly training workshop is too narrowly focussed on crop technologies.

A substantial training effort is needed to upgrade the technical and managerial/organizational skills, knowledge and qualifications of field and junior extension personnel. Some of the emerging areas of training are: Gender-related training, sustainable agriculture, participatory rural appraisal, participatory development, systems management, mobilisation of groups, small group dynamics, etc. It may be noted here that the capacity of Zonal Research Stations and State Agricultural Universities, which supply master trainers, is limited. Innovative and action-oriented NGOs may have comparative advantage in providing medium-term and short-term training. In the long-term, institutions like MANAGE should take lead to build a cadre of master trainers to fulfil the training needs. Simultaneously, substantial assistance is required in strengthening both the human resource capacity of extension, to discharge the role of human resource development and technology transfer, and the State Agricultural Universities that train extension personnel.

4.14 *Spreading the Cost of Agricultural Extension and Technology Transfer*

Currently public sector extension, private sector firms, NGOs, Cooperatives and semi-autonomous organizations are discharging the role of technology transfer in India. In certain commodities/regions/ sectors, technology has already shifted from being a "public good" to becoming a "private good." In this process, technology transfer costs are mostly paid by the farmers themselves, either through price of purchased inputs or through contracted consultancy services. In our classification scheme, high value commercial agriculture region is ripe for privatization. Similarly, a series of plant protection services in assured irrigated and high value crops in rainfed areas is in the process of being contracted out.

A publicly supported extension system is, however, essential to improve productivity, increase output and enhance food security in the majority of rainfed regions. In addition, along with concerned NGOs, public supported extension system should begin addressing issues of environmental degradation, and management of natural resources associated with sustainable agriculture.

Our recommendation on extension volunteers to be employed by farmers' organisation is likely to reduce budgetary burden on the exchequer. Another way for resource poor farmers to lessen the unit cost of extension services and for the government to increase its level of extension coverage is for extension to assume institution building role. It assists farmers to get organised into functional farmer organisations. This will accelerate farmer-to-farmer diffusion of new technology, increase accountability and responsiveness of extension and research, and increase the efficiency of extension in reaching a larger majority of its intended clientele.

Experimentation on cost-sharing extension also needs to be designed and implemented through the farmers' organization. The farmers' organization can identify and hire, from educated unemployed youth, potential extension agents, under mutually negotiated terms and conditions of employment and establish norms of performance evaluation and rewards. The extension, on the other hand can cover a fixed proportion of total costs which taper off once the experimentation is found to be viable and self-sustaining. More important role for extension, however, would be that of holding hand operation and providing backstop support in terms of creating opportunities for experiential learning, on the job training, publications, audio visual led distance learning and related support that is amenable for a centralized operation bacause of economies of scale. Also the extension would foster networking relationship amongst such locally appointed extension volunteers as well as between the extension volunteers and zonal research station researchers.

4.15 *Collaboration between Public Sector Extension/Research System and NGOs*

Non-government organizations (NGOs) have shown increasing commitment to embrace participatory approaches to agricultural technology generation and dissemination, specially in rainfed regions. NGOs have demonstrated comparative advantage in putting to practice farming systems approach to technology generation; technological and resources management innovations; participatory dissemination methods, training methodology, and building farmer organisations (Bebbington and Farrington, 1992). Public sector extension/research systems have been lacking these skills. While the NGOs have limitations like small size, focus on limited geographical area, limited search capacity and short-term project funding. It would appear that public extension/research systems can work together with NGOs in a complementary manner.

Training on participatory approaches, collaborative technology generation and dissemination on problems of common pool resource characteristics, collaborative programme to provide hands on experience for public sector extension on institution building, and networking relationships for NGOs to avail advantages of exhaustive and systematic search capacities of public research extension system, obtain access to public research/extension systems' repertoire of technology. Simulta-neously, the public sector extension can access NGOs software designs. These are some of the feasible areas of interdependent actions that would have mutual advantages. Besides, other potential roles for NGOs have also been identified under earlier recommendations.

5.0 Alternative Options

The study team has generated alternate options of agricultural extension and research for different agricultural situations. The options are summarized in matrix-form in Tables 2 and 3.

TABLE 2: **Salient Features of the Distinctive Extension Approaches for the Different Agricultural Situations in the Country**

Salient Features	Assured Irrigation System	Rainfed Farming System	High Value Commercial Crop System
1	2	3	4
1. Technology requirement	Specialised	Diverse	Highly specialised
2. Extension Approach (Dexter 1986)	Promotional (Agency decides on basis of views and needs of farmers)	Participative (Partnership between Agency & farmers at all stages)	Consultancy (Advice on request)
3. Focus on Extension	Crop based system	Farming system Based	Crop/tree based system
4. Program Content	Standardised Messages	Flexible Messages	Standardised Messages
5. Services	Production based	Resource based	Integrated Services (Post Harvest)
6. Clientele Extent to which to public sector has taken initiative in participatory approaches	Moderate	High	Low
Extension Volunteers	Not Required	Necessary	Not Required
Contact Farmers	Not Required	Not Required	Not Required
Gender Differentiation	Low	High	Low
7. Role of Extension Personnel	Advisory	Advisory Enabling Convening Consultancy and Tour Operator (Robert Chambers)	Advisory and Consultancy

(Contd.)

TABLE 2: *(Contd.)*

1	2	3	4
Deployment	At Block Level (As a team)	At village level	In production Areas
8. Area in which extn. workers require skills	Technological and problem Diagnosis	Technological and Behavioural	Technological
9. Extent to which training needs are currently being met	Moderate	Low	Moderate/High
10. Media Support	Modern Audio Visual and Print Media Support	Low Cost Indigenous Audio-Visual Aids	Sophisticated Communication Technology
11. Role of NGOs	Negligible	Important role in (a) Creating farmers' organisation (b) Training of farmers (c) Training of Extension volunteers (d) Training of Extension workers, especially in behavioural sciences and group dynamics (e) Drawing attention to systems based and "alternative" issues for research	Negligible
12. Cost-Sharing	Desirable	To be experimented	Must.

TABLE 3: **Distinctive Features of Proposed Agricultural Research in Different Agricultural Situations**

S. No.	Distinctive Features	Assured Irrigated Area	Rainfed Area	High value Commercial Agriculture
	1	2	3	4
1.	Nature of problems	* Declining Yield	* Low productivity	* Weather-induced losses
		* Sustainability issues, nutrient mining, pathogen buildup, dynamic weed regime, draw down of acquifer	* Fragile resource base	* Insect, pest incidence
			* High risk	* Management problems
			* Complex interactions among enterprises	* Quality control
		* Inefficient resource use	* High run-off losses	
		* Inefficient cropping pattern	* Problem soils	
2.	Methodological issues			
	(i) Desirable characteristics of research process	* Farmers monitoring of sustainability issues	* Farmer-centered restrospective diagnostics	* Contractual through growers' associations
		* Data base on indicators of non-sustainability	* Data base on indicators of non-sustainability	* Problem solving
		* Exploratory research on representative sites	* Inventory of indigenous knowledge systems	* Exploring new opportunities
		* Long-term experiments at research stations	* Linking of formal and informal R&D	
		* Interactions among inputs and soil-plant relationship	* Discerning farmers risk—risk-coping strategies—both long-term and short-term	
		* Simulation exercises com-	* Farmer participatory approaches to tech-	

(Contd.)

TABLE 3: *(Contd.)*

1	2	3	4
	modity based systems research * Research on common pool resource and governance * Explore new opportunities	nology generation and dissemination * Resource-based systems research * Research on Equitable use of common pool resources * Explore new opportunities	
(ii) Who designs* the experiments?	Researcher (increasingly researcher would have to incorporate farmers' understanding of sustainability)	Participatory Farmer, researcher, extensionists as equal partners	Commissioned Researcher designed on grower identified problems
(iii) Who implements the experiments?	Researcher+ Farmer	Depending on nature of problem Farmer alone Farmer and extensionist, Farmer, researcher and extensionist	Researcher
(iv) Monitoring & Evaluation	Researcher+ Farmer	Farmers' evaluation criteria incorporated by researcher and extensionist	Researcher and grower
(v) Methodological development	Researcher-led blending indigenous knowledge systems with formal methodology	Collaborative emphasis on blending indigenous knowledge systems with formal methodology	Researcher-led
(vi) Interaction between component and systems research	Interactive/ conflictive interactions, systems research feed forward ideas for controlled component research	Feed forward and feedback for conducting on-farm and on-station research	Stations based geared to address farmers' problems

(Contd.)

TABLE 3: *(Contd.)*

1	2	3	4
3. Institutional issues			
(i) Role of public sector	Medium to High	High	Low
(ii) Role of extension profes-sionals	* Help identify problems * Help mounting farmer moni-toring of sustain-ability issues * Help creating database trials and dissemi-nation	* Help conducting diagnostic * Help delineation of homogeneous agricultural situations and * Help identify repre-sentative sites * Help managing on-farm research * Help distilling lessons and feed-back * Demonstrations and dissemination * Human resources development and institution building	Promotional
(iii) Role of NGOs	* Minimal * Potential colla-borative role in governance of common pool resources insti-tutions.	* Important comple-mentary role in methodology development, methodology disse-mination, building farmers' organisation, helping to articulate farmers demands on research services.	* Minimal

ANNEXURE X

ITINERARY OF FIELD VISIT OF DR. D.K. SHARMA AND DR. CHARANJIT SINGH

Date	Place	Interactions
22.02.93	Nagpur	Shri R.N. Raut, Divisional Joint Director, Agr. - Nagpur Shri K.F. Kumbhare, Dy. Director of Agriculture Dr. B.S. Chimurkar, Extension Agroeconomist, Nagpur Dr. D.G. Bakwad, Principal Agriculture Officer, Latur
	Dhapewada	PRA village Dhapewada
23.02.93	Nagpur	Shri A.K. Niswada, Dy.Director, Agriculture T & V Personnel
24.02.93	Dhani	PRA in Dhani Village
25.02.93	Nagpur	Scientists of National Institute of Cibrus. Scientists of Central Institute of Cotton Research
26.02.93	Nagpur	Dr. R.R. Sinha, Director, Extension Education, PKV, Akola. Research & Extension Staff
01.03.93	Nevadia	PRA Village, Nevadia Shri M.L. Bhati, Joint Director Agriculture, Jodhpur PRA Village Atpada

(Contd.)

Date	Place	Interactions
02.03.93	Jodhpur	Meeting with extension staff. Scientists Zonal research station, Mandore
03.03.93	Jodhpur	Scientists of Central Arid Zone Research Institute, Jodhpur
08.03.93	Coimbatore	Dr. S. Chelliah, Director Research, TNAU Dr. V. Subramanyam, Director, Centre for Agriculture & Rural Development Studies & other Scientists.
09.03.93	Coimbatore	Directors and Professors of TNAU Extension staff department of Agriculture
10.03.93	Annamalai	PRA Village Annamalai
11.03.93	Pilliappan Palem	PRA Scientists of Research station Bhavani Sagar
12.03.93	Kappachi	PRA - Village Kappachi Nilgiri Hills Dr. K. Nanjan, Professor of Horticulture and other Scientists
13.03.93	Ooty	Scientists of TNAU stationed at Ooty and Extension staff
14.03.93	Conoor	Shri P. Swaminathan, United Planters Association of Southern India, KVK of UPASI
15.03.93	Madurai	Scientists of Agriculture College and Research Institute, Madurai
	Veerapandi	PRA of village Veerapandi
16.03.93	Kovilangulam	PRA of village Kovilangulam
	Arupukotai	Scientists of Dryland research station, Aruppukotai
17.03.93	Ekkakudi	PRA village Ekkakudi
18.03.93	Ramanathapuram	Extension staff of Ramanathapuram District and scientists of research station, Ramanathapuram

Date	Place	Interactions
19.03.93	Madras	Director of Agriculture, Tamil Nadu

ITINERARY OF FIELD VISIT OF DR. K.C. JOHN AND MR. C.S. RAJAN

Date	Place	Interactions
22.02.93	Calcutta	Mr. A Deb, IAS, Secretary (Agr.); Dr. D. Mukherjee, Director (Agr.); Dr. S. Biswas, Addl. Dir. (Research); Dr. M. Kr. Addl. Dir (T&V); Dr. Anil Mukherjee, FSRE Project Leader, BCKV, Kalyani; Dr. S.K. Samanti, Dy Director (Research), BCKV, Kalyani
23.02.93	Midnapore	Dr. A.K. Mandal, Jt. Director (Soil Conservation) and T & V Personnel
24.02.93	Midnapore	PRA in Sankoti Village; Research Personnel
25.02.93	Kardwip	PRA in Moinapara
26.02.93	Kardwip	T & V Personnel, Research Personnel from ZRS, Kardwip
28.02.93	Gauhati	Mr. P.B. Dhar, Addl. Dir (Hill)
01.03.93	Kahikuchi	Dr. U.C. Upadhyay, V.C., AAU; Dr. A.K. Roy, Dir (Research), AAU and Research Personnel of Horticulture Research Station, Kahikuchi
	Guwahati	T & V Personnel
02.03.93	Naogaon	PRA in Raidongia, Research Personnel of Zonal Research Station, Shillongani
03.03.93	Naogaon	T & V Personnel
04.03.93	Shillong	Secretary Agriculture, Director, Agriculture and Extension Personnel

(Contd.)

Date	Place	Interactions
		Dr. C.R. Patel, Scientist in-charge; Dr. Venugopal, Training coordinator for ICAR, North East Complex, Baipani, Shillong
05.03.93	Kahikuchi	Dr. Mohan, Asst. Director (Agr.); Chauwhans, Training Coordinator Horticultural Research Station, Kahikuchi
09.03.93	Lucknow	Mr. V.K. Agarwal, IAS, Sec. (Agr.); Dr. V.W. Ambekar, Director (Agr.); Dr. V.K. Singhal, Addl Dir (T&V) and T&V Personnel from Kanpur and Lucknow Division
10.3.93	Lucknow	Dr. V.W. Ambekar, Director (Agrl) Dr. R.K. Singh, Director (Research) Dr. I.S. Singh, Professor (Hort.) NDUAT, Faizabad Dr. Warsi, Director (Research) Dr. J.P. Yadav, Director (Extn.) CSUA & T, Kanpur Dr. D.C. Gupta, Secretary, U.P. Council of Agr. Research
11.03.93	Pantnagar	Dr. H.G. Singh, V.C. Dr. M.P. Singh, Director (Extn.) Dr. Vikram Singh, Joint Director (Research) and Research Personnel at Pantnagar
12.03.93	Pantnagar	PRA in Shantipuri Village, T&V Personnel from Bareilly and Moradabad Division Dr. S.C. Gupta, Associate Director (Extn.) GBPU, A&T, Pantnagar
13.03.93	Nainital	Personnel from Directorate of Agriculture, Hill Division

Date	Place	Interactions
16.03.93	Ranichauri	PRA in Arakot, Research Personnel from Zonal Research Station and KVK, Ranichauri
18.03.93	Meerut	PRA in Datawli, T&V Personnel from Meerut Division
	Modipuram	Research Personnel from Zonal Research Station, Modipuram

BIBLIOGRAPHY

Acharya, S., 1988. "Agricultural Wages in India: Trends and Regional Variations". Tata Institute of Social Sciences, Bombay, Mimeo.

Bebbington, A. and Farrington, J., 1992. "The Scope for NGO–Government Interactions in Agricultural Technology Development: An International Overview." Agricultural Research and Extension Network Paper No. 33, London, ODI.

Benor, D. and Harrison, J.Q., 1977. "Agricultural Extension". The Training and Visit System. Washington, D.C., World Bank.

Benor, D. and Baxter, M., 1984. Training and Visit Extension. Washington D.C., World Bank.

Biggs, S.D., 1989. "A Multiple Source of Innovation Model of Agricultural Research and Technology Promotion. Agricultural Research and Extension Network Paper No. 6, London, ODI.

Biggs, S.D., 1989. "Resource – Poor Farmer Participation in Research: A Synthesis of Experience from Nine National Agricultural Research Systems. OFCOR Comparative Study Paper No. 3, The Hague, ISNAR.

Biggs, S.D. and Farrington, J., 1991. "Agricultural Research and the Rural Poor: A Review of Social Science Analysis" Report IDRC-280e, Ottawa, International Development Research Centre.

Bredero, T.J., 1988. "Status Reports: A Basis for the Planning of Agricultural Research, Extension and Other Rural Development Activities in India. Agricultural Research and Extension Network Discussion Paper No. 26, London, ODI.

Caroll, T., 1992. *Intermediary NGOs: The Supporting Link in Grassroots Development.* West Hartford, Kumarian Press.

Chambers, R. and Ghildyal, B.P., 1985. "Agricultural Research for Resource—Poor Farmers: The Farmer-First and Last Model". *Agricultural Administration,* 20, pp. 1-30.

Chambers, R., Pacey A. and Thrupp L.A. (eds.), 1989. *Farmer First: Farmer Innovation and Agricultural Research,* London, Intermediate Technology Publications.

Farrington, J. and Martin. A., 1988. Farmer Participation in Agricultural Research: A Review of Concepts and Practices, Occasional Paper No. 9, London, ODI.

Feder, G., and Slade, R., 1986. "Institutional Reform in India: The Case of Agricultural Extension." Washington, D.C., The World Bank. Mimeo.

Fernandez, A.P , 1991. "MYRADA", Participatory Rural Appraisal and Hearing Methods". Bangalore, MYRADA. Mimeo.

Government of India, 1988. "Report of the Screening Committee on the Findings of the World Bank Mission's Review of the Agricultural Extension Sector in India", Ministry of Agriculture, Department of Agriculture and Cooperation (Extension Division), New Delhi, Mimeo.

Gupta, A.K., 1987. "Technology for Dry Farming: How the Scientists, Students and Farmers View the Challenge?" Working Paper 708, Ahmedabad, Indian Institute of Management.

ICAR, 1988. "Report of the ICAR Review Committee," ICAR, New Delhi.

ICAR, 1989. "National Agricultural Research Project-II: Mid-term Review" Project Unit (NARP), ICAR, New Delhi.

John, K.C., 1990. "Farming Systems Research Programme in Eastern India: An Overview", New Delhi, The Ford Foundation, Mimeo.

John, K.C., 1991. "Rice-based Farming Systems Research—A New Approach" in Proceedings of International Workshop on Frontiers of Rice Research. Directorate of Rice Research, Hyderabad.

Mandal, B.; Das C. and Chakraborty, S., 1991. "Rama Krishna Mission: Research, Extension and Training—Report , Rama Krishna Mission, Mimeo.

Merrill – Sands, D. and Mc Allister, J., 1988. Strengthening the Integration of On-farm Client-Oriented Research and Experiment Station Research in National Agricultural Research Systems (NARS): Management Lessons from Nine Country Case Studies. OFCOR Comparative Study Paper No. 1. The Hague, ISNAR.

Merrill – Sands, D. and Kaimowitz, D., 1991. The Technology Triangle: Linking Farmers, Technology Transfer Agents and Agricultural Researcher. The Hague, ISNAR.

Mishra, P.R., John K.C. and Bottrall, A.F., 1992. "Management of Property Rights Regime: An Empirical Exploration of Equity and Sustainability," Paper at the Conference of the International

Association for the Study of Common Property, Washington, D.C.

Randhawa, N.S., 1987. "Agricultural Research in India: An Overview of its Organisations, Management and Operations". Research and Technology Paper No. 3, Rome, FAO.

Sanghi, N.K, Kerr, J.M., Qayum, M.A., Rao M. and Reddy, R., 1992. "Blending of Formal and Informal Knowledge in Dryland Agriculture: 3 case studies from Andhra Pradesh, India". Paper at Beyond Farmer First: Rural People's Knowledge Workshop, Sussex, IDS.

Saran, S., John K.C. and Kumar, A., 1993. "Integrated Farming Systems Methodology for High-Risk Ecological Zones." *Journal of the Asian Farming Systems Association,* Vol. 2 (in press).

Satish, S.; Vardhan, T.T.P.S. and Farrington, J., 1991. "Integrated Pest Management in Castor: A Participatory Programme Involving NGOs, Public Sector Research and Extension Agencies and Farmers in Andhra Pradesh". Paper presented at the Asian Regional Workshop NGOs, Natural Resources Management and Linkages with the Public Sector, Hyderabad, India.

Shaner, W.W., Philipp, P.F. and Schmehl, W.R., 1982. *Farming System Research and Development: Guidelines for Developing Countries*—Colorado, Westview Press.

Vasimalai, M., 1991. "Livelihood Promotion Technology for the Rural Poor—The Case of PRADAN". Paper presented at the Asian Regional Workshop NGOs, Natural Resources Management and Linkages with the Public Sector, Hyderabad, India.

World Bank, 1988. "India: Agricultural Extension Sector Review," Washington, D.C., Mimeo.

INDEX

Action for World Solidiarity (AWS), 58-59, 172, 176
Aga Khan Rural Support Programme (AKRSP), Gujarat, 58, 60, 105-06, 170, 172, 176, 192, 201-05
Agricultural Administration (Research and Extension) Network, London, 17
agricultural extension, 45-46, 85, 96-98; and research, policy, 65-68, 242-43; — zonation for, concept of, 26-27; service reforms in India, 213-14; — cost sharing, 229; — farmers participation, 65-68, 242-43; — need for innovations, 214-15
Agricultural Knowledge Information System, 63-65
Agricultural Products Cess Fund, 250
agricultural research, 84-85
Agricultural Research and Education Department (DARE), 40
Agricultural Technology Development and Dissemination System, 40*ff*
Agriculture Directorate, 48-49
AGRITEX, Zimbabwe, 152-53
Agro-Climatic Regional Planning Unit, 26
Agronomy Institute, Research and Specialist Services Department, Zimbabwe, 156
Ahmadu Bello University, Nigeria, 151
All India Coordinated Research Projects (AICRP), 41, 59, 78
Andhra Pradesh: Agriculture Department, 51-52; irrigation, 51-52; National Watershed Development Programme (NWDP), 48, 170; National Watershed Development Programme for Rainfed Agriculture (NWDPRA), 52; non-government organisations' role, 59; rainfall, 34; Training & Visit System, 48
Animal Husbandry Department, Rajasthan, 100
acquaculture based farming systems, 35-36
Ardhanareeswaran Committee *see* Screening Committee on Findings of the World Bank Mission's Review of the Agriculture Extension Sector in India
Ardhanareeswaran, K.N., 47
Assam: Agriculture Department, 54-55; Field Management Committee, 54-58, 102, 105; public agriculture extension, 54-55; rainfall, 34
Auroville Greenwork Resources Centre, 176

Bangladesh: Agriculture Extension Department, 155; Extension and Research Project, 139-40, 155-56; Innovative Farmers Workshop, 139-40; internal review workshops, 155-56
BHAGIRATH, 77
Bharat Starch Company, Haryana, 175
Bharatiya Agro-Industries Foundation (BAIF), 58-59, 170, 172, 176, 181-86
Bihar: arrest of environmental degradation, 59-60; rainfall, 34
Bolivia: Centro de Investication Agricola Tropical (CIAT), 148; non-government organisations'

role, 148-49
British Tropical Agricultural Mission, 148
Bureau of Plant Industry, Philippines, 145

CAPART, 60, 192
CIMMYT, 18
Centrally Sponsored Special Food Programme, 51
Centro International de Agricultura Tropical (CIAT), Colombia, 160-62
Centro de Investication Agricola Tropical (CIAT), Bolivia, 148
Client driven extension research system, 244
Coffee Growers Federation, Colombia, 149
Cogmill India, Karnataka, 175
Colombia: Caqueza development project, 162; Centro International de Agricultura Tropical (CIAT), Colombia, 160-62; Coffee Growers Federation, 149; communication plans, 159-60; Institute Colomniano Agropecurario, 140; International Research and Development Programme, 137-38, 144; Llanes Orientales, 160; mixed farming, 137-38; policy intervention, 143-44; research and extension, 149-50; status differences, 162-63
Commercial agriculture, 37-39
Communication plans, Colombia, 159-60
Communication support: role in creating farmer awareness, 80-81
Cornell University, Philippines, 145
Cost reduction strategies, 82-83
Cost sharing, 229
Costa Rica: Agriculture and Livestock Ministry, 153; research and extension, failure, 153-54

DARE *see* Agricultural Research and Education Department
Deccan Development Society, 58, 172, 176
Dryland Development Board of Karnataka, 102

Eastern India Farming Systems Research Extension Network, 174
Electronics Department, 47
extension services, role, 171

Farm Extension Volunteers (FEV), 67-68, 75, 247-49
farmer participatory approaches, 245-47
Field Management Committee (F.M.C.), 54-58, 102, 105
farmer participatory extension and research, policy support, 83-84
farmer participatory process, management, 73-80
farming systems research extension (FSRE), 17
Ford Foundation, 192

G.V.K. Rao Committee *see* Indian Council for Agriculture Research; Review Committee
Gambia: Farmer Innovation and Technology Test Programme, 147-48; non-government organisation, 147-48; tailoring linkage strategies, 147-48
Gopals: the community livestock volunteers, Rajasthan, 106
Gram Vikas Mandals, Gujarat, 105
Grape Growers' Association in Maharashtra, 109, 175
GROFED *see* Gujarat Cooperative Oilseeds Growers Federation Limited
group approaches, 103-06
group formation, guiding principles, 66-67, 226-29
'Group Trek', Nepal, 154-55
Guatemala: Agriculture Ministry, 163; Instituto de Ciencay Technologic Agricolas (ICTA), 141-42; integrated research extension project, 141-43
Gujarat: Fisheries Development Department, 185; Minor Irrigation Department, 182-86; Padhar Project, 182-86; pastoral system,

240; rainfall, 34; Tribal Development Department, 182-85
Gujarat Agriculture University, 189
Gujarat Cooperative Oilseeds Growers Federation Limited (GROFED), 170, 172, 186-91

High Yielding Varietal Programme, 213
Himachal Pradesh: horticulture based system, 33; rainfed areas, 33; Training and visit System, 48
Hindustan Fertilizer Corporation, 175
horticulture-based farming system of hills, 35

IRRI, 18
improved training methods, 60-62
India: agricultural situation analysis, 27-30, 236-37; extension services, reforms in, 213-14; —alternate roles for, 223-24; —cost sharing, 229; —farmers' participation, 174-75; —need for innovations, 214-15; —non-government organisations' role, 176, 229-30; —research extension linkage, 230-31; farming systems, 34-37; growers' associations, 175; rainfall and irrigated areas, 30-33; rainfed areas, 33-34, 240-41
Indian Council for Agriculture, 26, 40-41, 78, 174-75, 182, 250; Review Committee, 41
Indian Tobacco Company, 171-72; Agro Business Division, 108
Indo-British Rainfed Farming Project, 195-201
Innovative Farmers' Workshops, Bangladesh, 139-40
Institute Senegalais de Recherches Agricolas (ISRA), 146
Instituto Colomniano Agropecuario (ICA), Colombia, 140-41
Instituto de Ciencay Technologia Agricolas (ICTA), Guatemala, 141-42
Integrated Pest Management, 67
Integrated Rural Development Programme, 137-38
Integrated Tribal Development Programme, 192
Intensive Agriculture Development Programme, 51
Intensive Area Agriculture Programme, 213
Intensive Area Development Programme, 213
International Agricultural Research Centre, 144, 231, 235
International Institute of Rural Reconstruction, Philippines, 176
International Service for National Agricultural Research, 17
irrigated areas, options for, 84-99

Jagawat, 193
Joint Extension Research Management Board at zonal level, proposed, 79
Joint Research and Extension Management Committee, 111, 116-17, 121

Kenya: group approach in processing of coffee, 105
Kerala: public agriculture extension, 55
Kisan Sewa Kendras, Uttar Pradesh, 102
KRIBHCO *see* Krishak Bharat Cooperative Limited
Krishak Bharat Cooperative Limited, 170-72, 175, 195-98
Krishi Vigyan Kendras, 45, 231; Banswara, 198

Lakhvad Society, 188
Lift Irrigation Project, Shankarpura, 192-94
Lift Irrigation Scheme, 204-05
Limkheda (North) cluster: subsistence form of agriculture, 198-201

Maize Growers Association, India, 175
MYRADA, watershed development project, Gulbarga, 58, 170, 172, 176
Malawi: group approach, 105
mixed farming, 137-38

N.M. Sadguru Water and Development

Foundation, Dahod, 170, 172, 191-95, 201
National Agricultural Extension Projects (NAEP), 15, 18
National Agricultural Research Project (NARP), 15, 18, 26, 42-44, 132, 134, 213
National Agricultural Research System (NARS), 17, 40, 73, 174
National Bureau of Soil Survey and Land Use Planning (NBSSLUP), 26
National Coconut Development Programme, Tanzania, 163-64
National Dairy Development Board (NDDB), 174, 186, 188
National Extension Services, 213
National Institute of Agricultural Extension Management (MANAGE), 170, 222-23, 233*fn*, 251
National Oilseeds Development Programme, 48-49
National Pulse Production Programme, 49
National Watershed Development Programme (NWDP), 48, 170
National Watershed Development Programme for Rainfed Agriculture (NWDPRA), 52
Nepal: Farming Systems Research and Development Division, 155; Group Trek, 154-55; joint planning, 154-55
Nigeria: Agriculture Ministry, 151; Ahmadu Bello University, 151; extension research liaison service, 151-52
Nithin Kribhco, 197
non-government organisations', role, 57-65, 69-70, 72-73, 78, 82, 84, 98, 105-06, 116, 122, 168-69, 172-73, 176-78, 217, 231, 245-46, 249, 251-53
NORAD, 192
Northern Philippines Root Crops Research and Training Centre, 145

Oilseed Technology Mission, 170
on-farm research, 141, 143, 155
Orissa: rainfed areas, 33-34
Overseas Development Administration, 196

Padhar project, Gujarat, 182-86
participatory agriculture development, 68-73, 76
participatory rural appraisal, 68, 76, 172, 197-98, 203, 235
pastoral based systems, 34-35
Patel, I.G., 202
Patel, M.D., 188
Pepsi International, 109
Philippines: Bureau of Plant Industry, 145; Cornell University, 145; donor involvement, 144-45; Northern Philippines Root Crops Research Institute and Training Centre, 145
Piludora Society, 188
Planning Commission, 26
PRADHAN, 59, 176

rainfed crop based farming system, 34
rainfed farming system, 224
rainfed situations, 99-103
Rajasthan: Agricultural Directorate, 49-51, 55; Animal Husbandry Department, 100; pastoral systems, 240; rainfed areas, 33; rainfall, 34; Training and Visit System and Special Schemes, 49-51; united funds schemes, 77; village extension workers, 50-51
Rajedia Project, 192-94
Ramakrishna Mission, Narendrapur, 58-60, 170, 172, 176, 243
Research and Extension Coordination Committee, 56-57, 231
research-extension integration, 141-43
research-extension liaison positions, Zambia, 150
research extension linkage, India, 230-31
Rupa Bhai, 193
Rural Development and Agriculture Department (DRDA), 192, 205

Sadguru Seva *see* N.M. Sadguru Water and Development Foundation
Sanghi, N.K., 53
Screening Committee on the Findings of World Bank, 47
Senegal: donor driven reorganisation, failure of, 146-47; Institute Senegalais de Recherches Agricoles, (ISRA), 146; Production Systems Department, 146-47; reorganisation of linkages, 146
Seva Bharati, 58
Society for Hill Resource Management School, India, 59, 176
Society for Rural Industries, 59
Somalia: group approach in groundnut crop, 105
Special Crops Production Programme, 213
Special Food Production Programme, 49
Sri Aurobindo Krishi Vigyan Kendra, 60
STANROSE, 192
State Agriculture Universities (SAU), 15, 17, 40-43, 82, 85, 132, 134, 174, 214, 231, 236, 251

Tanzania: National Coconut Development Programme (NCDP), 163; status difference, 163-64
Tarai Development Corporation (TDC), 48-49
Tribal Development Department, Gujarat, 182-85

United Planters Association of Southern India (UPASI), 109, 175
United States Agency for International Development (USAID), 144
Uttar Pradesh: Krishi Vigyan Kendra, 45, 102; rainfall, 34; Zonal Research Stations, 45

village extension workers (VEWs), 67, 75-77, 85, 100, 102, 106, 114, 117-23, 132, 190, 213, 223

West Bengal: crop based system, 33; group approach, 105; non-government organisations' role, 59; rainfed areas, 83; rainfall, 34; rural youth contribution, 107
World Bank, 106, 132, 146

Youth Clubs, West Bengal: farmer coverage, spread of technology, 105

Zambia: Adaptive Research Planning Team (ARPT), 159; Agriculture Ministry, 157; Provincial Research and Extension Committee, 157; research extension liaison implementation problems, 150, 157
Zimbabwe: AGRITEX, 152-53; Agronomy Institute, 156; Committee for on-Farm Research and Extension (COFRE), 152-53; policy intervention, 143; research extension collaboration, 156-57; Research and Specialist Services Department, 152, 156
Zonal Research and Extension Advisory Committees (ZREAC), 43-44, 57, 174, 231
Zonal Research Stations, 78-79, 82, 84, 111, 116-17, 121, 213, 231, 244, 249-51